FOREST SERVICE FIRE & AVIATION QUALIFICATIONS GUIDE

February 28, 2011

File Code: 5100   **Date:** February 14, 2011
Route To:

Subject: Change to FSH 5109.17, Issuance of Forest Service Fire and Aviation
Qualifications Guide

To: Regional Foresters, Station Directors, Area Director, Regional Fire Directors,
Regional Aviation Officers

Forest Service (FS) policy relating to Fire and Aviation Management Qualifications in the FSH 5109.17 is outdated due to the rapid changes in training courses and career development processes. This creates serious problems for FS personnel in meeting qualifications as well as succession planning.

Effective February 28, 2011, "The FS Fire and Aviation Management (FAM) Qualifications Guide" will replace Chapters 20, 30 & 40 of the FSH 5109.17 and will be referenced in the FSH 5109.17 Chapters Zero Code and 10 Code as policy. The Forest Service Fire and Aviation Qualifications Guide will be on the FS FAM web page at: http://www.fs.fed.us/fire/index.html with current requirements for FAM positions.

This guide will be updated annually in conjunction with the National Wildfire Coordinating Groups issuance of the PMS 310-1 Wildland Fire Qualification System Guide as well as when updates are issued to the PMS 310-1.

Maintaining this guide will be the responsibility of the Branch Chief, FAM Training, and an advisory council consisting of the Regional training officers, selected subject matter experts, and a union representative.

Please direct any questions about this update to Jill McCurdy, Branch Chief, Fire and Aviation Training, at jmccurdy@fs.fed.us or 208-387-5737.

/s/ *JAMES E. HUBBARD*
JAMES E. HUBBARD
Deputy Chief, State and Private Forestry

cc: Wm C Waterbury, Neal Hitchcock, Larry Sutton, Ron Hanks, Jill M McCurdy

America's Working Forests – Caring Every Day in Every Way   Printed on Recycled Paper

FOREST SERVICE FIRE AND AVIATION QUALIFICATIONS GUIDE

CHAPTER 2 PART 1 - QUALIFICATIONS AND CERTIFICATION
CERTIFICATION, DECERTIFICATION, FITNESS

Effective Date: February 28, 2011, Updated 6/10/2011

Updates:

See "Summary of Changes" document available at:
http://www.fs.fed.us/fire/publications/index.html#fsfaqg

Table of Contents

INDEX TO NWCG POSITIONS

2.0 - INTRODUCTION

This chapter sets out the interagency requirements outlined by the National Wildfire Coordinating Group (NWCG) in the Wildland Fire Qualification System Guide, PMS 310-1, additional interagency requirements outlined in guides referenced in Zero Code of the FSH 5109.17, and additional Forest Service requirements for incident and prescribed fire management positions.

In addition, standards established for technical specialist positions are included in this handbook.

The complete list of 4-letter position job code identifiers, which are approved by the NWCG Position Naming Board, are maintained on the Incident Qualifications and Certification System website: *http://iqcs.nwcg.gov/*.

Direction related to training courses, delivery, equivalency, and instructor qualifications can be found in chapter 3 of this Guide.

The National Fire and Aviation Executive Board developed the Interagency Fire Program Management Qualifications Standards and Guide (FSM 5108), which contains minimum qualification standards for fire and aviation management positions. Direction related to competencies for fire and aviation management positions is contained in chapter 4. Additional information can be found at *http://www.ifpm.nifc.gov/*.

2.1 - QUALIFICATION FOR INCIDENT AND PRESCRIBED FIRE POSITIONS

This section sets out the additional requirements for skills, training, and prerequisites for Forest Service employees for the incident and prescribed fire management positions. Qualifications for each position include the minimum acceptable levels of training, experience, physical fitness, and/or currency requirements (ch. 2) except where noted in this Guide.

Job descriptions and duties for these positions are listed in the NWCG Fireline Handbook, PMS 410-1, or in the interagency guides and agency directives listed in Zero Code of the FSH 5109.17.

1. The successful completion of a position task book during an appropriate number of evaluation assignments is the indicator of qualification. For positions that do not have a position task book, the Forest Qualification Review Committee shall review and recommend to the certifying official an individual's certification and qualification, based on objective factors such as performance evaluations and visual observation of performance of duties of the positions.

2. Personnel who have learned skills from sources outside the wildland fire management curriculum may not be required to complete specific NWCG courses in order to qualify in an incident and prescribed fire management position (sec. 3.2 - Course Equivalency).

3. Personnel currently employed by state agencies or other non-Federal cooperators (such as rural, county, and city fire departments) and hired as Administratively Determined (AD) personnel participating in cooperative fire management efforts, must be compliant with the minimum position requirements specified in the NWCG Wildland Fire Qualifications Systems Guide, PMS 310-1. Additional training requirements contained within this handbook do not apply to these individuals when they are hired,

since the Forest Service is hiring these individuals on a temporary basis to perform in positions where their agency has already certified them.

4. For contract fallers and equipment operators, hired under Emergency Equipment Rental Agreements (EERAs), designated for fireline operations, the hiring unit or hiring official is authorized to evaluate their knowledge, skills, abilities and associated certifications or past performance records. Individuals who serve in these positions and have existing contracts which outline specific knowledge, skills, and abilities may utilize these documents as evidence. If determined sufficient, these individuals may be utilized as a resource on local incidents, when they can be supervised by falling bosses, dozer bosses, or other similar certified supervisors.

When employed specifically for their occupational skills, physical fitness testing and course requirements for these positions do not apply (FSM 5134.2). Pursuant to section C of PMS 310-1, "Agencies shall not certify private contractors except where formal agreements are in place". The Forest Service obligation is to monitor and evaluate the performance and safety of these technical specialists.

2.2 - CERTIFICATION

1. All regular Forest Service personnel and cooperators employed by the Forest Service under the Pay Plan for Emergency Workers (Interagency Incident Business Management Handbook, PMS 902-1) and assigned Incident Command System (ICS) positions must be judged to be technically and physically qualified to fill their positions.

2. Qualification for a position in the NIMS/ICS organization depends on proven ability (sec. 2.1). Training, experience, and physical fitness are prerequisites for qualification. However, certification to hold an ICS or prescribed fire position is determined through evaluation of performance as a trainee in the target position (if required), or in a prerequisite assignment.

The Washington Office, Regional or Forest Qualification Review Committees shall determine when the individual is prepared to advance to the next higher level and make recommendations to the certifying official. Documentation of the recommendation and decision shall be placed in the employee's master record file (sec. 2.21).

3. Each employee's incident and prescribed fire position qualifications (including physical fitness and experience) must be re-evaluated annually by the certifying official to determine certification, recertification or decertification and a new Incident Qualification Card must be issued. This includes all Forest Service employees as well as Administratively Determined (AD) employees certified by the Fire Program Management Staff Officer.

4. Employees transferring to the Forest Service from another NWCG agency must be re-evaluated by the certifying official to determine certification.

5. The Fire Program Management Staff Officer should ensure that a system is in place to establish priorities for training and currency assignments.

6. A Qualification Review Committee shall be established on each unit (Washington Office, regional office, or forest) to review and recommend certification of personnel to

the certifying official. At a minimum, the committee shall include the Fire Management Staff Officer, a line officer representative, Incident Qualifications and Certification System (IQCS) Administrator, and a representative from the National Federation of Federal Employees (NFFE) or other appropriate Forest Service union official, as well as representatives knowledgeable of the unit's personnel. Deliberations, rationale and decisions must be documented as appropriate to establish criteria and provide background for employee performance enhancement planning.

    a. The Washington Office Review Committee (WQRC) shall review all individuals possessing Area Command and Type 1 Command and General Staff position qualifications, who are assigned to the Washington Office.

    b. Regional Qualification Review Committees (RQRCs) shall review all individuals possessing Area Command and Type 1 Command and General Staff position qualifications who are assigned in that Region.

    c. Forest Qualification Review Committees (FQRCs) shall review all individuals possessing Type 2, or lower, position qualifications.

7. If an individual is serving in the designated 3 year Task Book completion period and an amendment to FSFAQG is issued, the position standards, contained in the amendment apply.

8. If amendments to FSFAQG are issued which identify additional required training for positions, such training will not be required for individuals qualified and current for those positions at the time of amendment. There may be exceptions, examples include

Homeland Security Presidential Declaration #5 (HSPD-5) which required successful

completion of IS-700 for all employees, whether or not they were previously qualified.

Recommendations from Accident/Incident Abatement Plans may also require training for

all employees, whether or not they were previously qualified. An example is the Time

Pressure Simulation Assessment for all Incident Commanders, Type 3. Field units will

be notified when additional training is required through a letter from the Deputy Chief of

the training requirement and who it applies too.

9. Specific training may be required by the Occupational Safety and Health Act

regarding hazardous waste operations during emergency response and may be

coordinated by Occupational Safety Health Administration (OSHA).

2.21 - Annual Fireline Safety Refresher

. Detailed information on refresher training can be found in chapter 3, 3.13.

2.22 - Record Keeping

Beginning with the March 26, 1990, amendment to FSH 5109.17, the Forest Service requires that

certification records supporting qualifications on the employee's Incident Qualifications Card be

maintained. The forest fire program management staff officer on each Forest shall ensure that

certification records are maintained.

1. A file folder in paper copy format must be maintained for each employee for records

related to certification in fire and aviation management positions. These records must be

centrally located on the unit and readily accessible to the fire program manager. The fire program manager shall determine the centralized location of the records and provide direction to the units.

2. The contents of these folders shall include:

a. All relevant evidence of course completion related to position qualifications. Relevant evidence may consist of training certificates, copies of course completion rosters and/or completed Request for Training (SF-182) forms, or other third party validation that the training has been successfully completed. Documents from the course lead instructor or course coordinator, certifying that the employee successfully completed the training, are acceptable.

Only the lead instructor, course coordinator, or an individual who can validate the responder's successful completion of the training can provide third party evidence to substantiate course completion. An individual who also attended the course is not acceptable third party evidence.

b. Individual Performance Rating, Form ICS 226. Prior to the implementation of Position Task Books in February 1994, the Individual Performance Rating, Form ICS 226, served as evidence that the employee satisfactorily performed in a position.

If an Individual Performance Rating recommending the individual for certification cannot be located for the periods from March 1990 through February 1994, verification can be accomplished with a letter or e-mail from the final evaluator validating:

(1) A recommendation that the individual be considered for certification.

(2) A list of the incident(s) and date(s) when the final evaluation and previously recommended certification was completed.

c. Position task book verification (the inside front cover of task book, showing recommending final evaluator and Certifying Official's signatures and dates). If no copy of a qualifying position task book verification page can be located, verification may be accomplished with a letter or e-mail from the final evaluator validating:

(1) Satisfactory completion of all tasks.

(2) A recommendation that the individual be considered for certification.

(3) A list of the incident(s) and date(s) when the final evaluation and previously recommended certification was completed.

d. Other Documented Evidence of Continued Certification. When course completion and position task book verification evidence is missing, the certifying officials must also document that the employee was eligible for certification and their decision to continue to certify the employee in the position. In each case, the certifying officials must document their rationale to retain the employee's qualification.

e. Yearly updated Incident Qualifications and Certification System Responder Master Record (RPTC028) from IQCS.

3. Decertification. Decertification of an individual's ability to perform is the responsibility of the employing line officer at the Washington Office, region, forest, or district level. Decertification records shall be maintained in the employee development folder (EDF), which is maintained by employee's work supervisor.

4. Record Keepers must take the IQCS course. Everyone that takes the IQCS course must sign a "Statement of Responsibility" and "IQCS Rules of Behavior". The Statement of Responsibility form identifies "GENERAL BUSINESS PRACTICES - INCIDENT QUALIFICATIONS AND CERTIFICATION SYSTEM (IQCS)". These forms commit them to uphold security requirements within IQCS. This includes commitment to adequately protect and not share individual user IDs and passwords. In support of these agreements, all those Agency employees that have authorized access to IQCS shall maintain strict access/security protocol. If they are found in violation of the protocol, their access will be suspended and the following mitigation(s) will apply:

First Violation: The supervisor, certifying official or staff officer shall determine the violation causal factor(s) and develop a mitigation plan of action. Causal factors may include, but not be limited to, inordinate volume of data entry; insufficient authorized staff; inappropriately targeted account holder; insufficient protection of user ID and password; pressure to share access, etc. The plan of action will require total mitigation by the represented unit. Individual access will be reinstated after the plan is completed and submitted to and approved by the Agency IQCS Representative.

Second Violation: Re-evaluation of first violation mitigation. Removal of access until the user completes a review of IQCS "Statement of Responsibility" and "IQCS Rules of Behavior" associated with security responsibilities.

Third Violation: Re-evaluation of first violation mitigation. Removal of access until the user completes the next available Security Training in AgLearn. Individual access will be reinstated with the submission of Certificate of Training to the Agency

IQCS Representative.

Fourth Violation: Permanent removal.

2.23 - Position Task Books

Position Task Books (PTB) identify all critical tasks required to perform the job for most

standard incident management and prescribed fire positions. Position Task Book

Responsibilities are contained within the Wildland Fire Qualification System Guide, PMS 310-1.

1. An individual may not have more than six active position task books at one time. No

    more than two of the six allowed position task books may be in a single functional area,

    including prescribed fire positions. The functional areas include:

    a. Command and General Staff

    b. Finance

    c. Logistics

    d. Operations

    e. Air Operations

    f. Planning

    g. Prescribed Fire.

h. Incident Support and Associated Activities (examples includes Expanded Dispatch or Wildland Fire Investigation).

2. In a few instances, agency or interagency position task books have been developed for positions outside those identified in PMS 310-1. Where these PTBs exist and have a Forest Service Logo on the front page, they shall be used to complete performance evaluations. The PTBs will be reviewed for qualification using the same procedures as those sponsored by NWCG.

3. For positions that do not have a position task book, the Forest Qualification Review Committee shall review and recommend to the certifying official an individual's certification and qualification, based on objective factors such as performance evaluations and visual observation of satisfactory performance of position duties.

2.24 - Currency Requirements

1. Currency requirements are contained in PMS 310-1. Currency requirements for air operations positions may also be met by performing on a day-to-day basis or on special projects, such as aerial spraying, search and rescue, and aerial ignition on prescribed burns. The day-to-day operational standards for these positions are governed by the procedures outlined in the Interagency Helicopter Operations Guide (IHOG).

Currency requirements for Security Manager (SECM) and for Security Specialist Type 1 (SEC1) may also be met by employment in good standing as a Forest Service Criminal Investigator or Law Enforcement Officer performing security and law enforcement work.

The day-to-day operational standards for these positions are governed by the procedures outlined in FSM 5300 and FSH 5309.11 and must be followed in order to meet the currency standard. Such law enforcement personnel meeting these currency requirements shall be identified as provided for in FSH 5109.17, Zero Code, Section 04.1-2 and Section 04.3-3.

2. Position experience is considered as qualifying only if the individual has previously met all prerequisite requirements for the position assignment.

3. If a position currency on an individual's Qualifications Card expires while they are performing that position on an assignment, the individual shall be allowed to complete the assignment. However, they shall not be reassigned in any expired position by the incident or home unit until they have re-established currency through the FQRC for that position.

2.25 - Recertification

Management evaluation of an individual's competency is key to recertification where qualifications have expired.

1. If currency has lapsed, the individual shall revert to the trainee level in the position for which currency has lapsed and shall be issued a position task book for the position, complete on-the-job-training if needed, complete any additional required training courses which have been added to the position for which they are attempting to recertify and requalify in the related position performance assignment.

2. Incident Commanders, Type 3s, who have lapsed currency shall complete an approved Time Pressure Simulation Assessment (TPSA) prior to being recertified.

2.26 - Decertification

The decertification procedures in this section are intended to ensure safe and effective individual performance in assigned ICS, Wildfire Skill, Technical Specialist, and Prescribed Fire Skill positions. These procedures are also intended to provide supervisors and managers with an additional mechanism to ensure employee safety.

Decertification is the process of removing or reducing an individual's fire suppression and/or prescribed fire management position(s) qualifications. Decertification is not an adverse action; an employee may be recertified according to the procedures set out in section 2.25.

1. <u>Performance Issues Outside Fire Suppression and Prescribed Fire Management</u>. Different procedures are utilized to address individual performance issues in areas other than fire suppression and prescribed fire management. Refer to Agency Human Resources Policy.

2. <u>Causes for Decertification</u>. There are three causes for losing certification (decertification):

a. An employee who currently holds a certification does not meet the currency requirements as specified by this Guide.

b. An employee voluntarily surrenders the employee's certification of qualifications or requests to be qualified at a lower level of responsibility.

c. As an individual or a member of a crew, incident management team, or prescribed fire team, an employee performs actions that violate recognized standard operational procedures or identified safety procedures that are determined to have been instrumental in the endangerment of fire management personnel or the public.

Examples of instances that may warrant decertification include:

(1) Deliberately disregarding identified safe practices.

(2) Taking insubordinate actions that lead to unsafe conditions.

(3) Intentionally misrepresenting fire qualifications.

(4) Ignoring prescriptive parameters identified in approved burn plans.

3. Performance Evaluation and Documentation. Performance of personnel shall be evaluated on each incident. The Forest Service shall utilize the adopted interagency team and individual performance rating forms (or recognized equivalent) when evaluating the performance of individuals.

All actions that violate established safety procedures shall be documented; associated deficient performance evaluations must also be completed. Performance reviews, especially those that trigger consideration of decertification, shall be coordinated and tracked.

4. Responsibility for Performance Evaluation and Decertification.

a. Incident Commander. The incident commander and local unit manager are responsible for providing oversight of the initial performance review process. Inherent within the authority delegated to all incident commanders is the responsibility to relieve from assignment and demobilize any personnel for safety violations. Incident Commanders, however, do not have the authority to decertify individuals. Incident Commanders are responsible for providing documented reasons for relieving an individual, forwarding the information to the individual's home unit, and including a copy of the individual's performance rating in the documentation package.

b. Forest Fire Program Management Staff Officer. The forest fire program management staff officer at the home unit is responsible for initiating an administrative review to determine if decertification is appropriate.

Any decision to decertify an individual should include a determination of whether remedial actions are appropriate to recertify the individual and a description of the recommended remedial actions.

During an evaluation of decertification, individual qualifications may be temporarily suspended. Judgments about qualifications can be made through expert mentoring, independent assessment, or the line officer's judgment relating to the individual's performance capabilities.

c. Washington Office, Regional and Forest Qualification Review Committee. The Washington Office, Regional Office, and Forest Qualification review committees are a key component in the certification and decertification of individuals. Qualification review committees should operate according to procedures delineated in other sections of this handbook and in FSM 5120.

Qualification Review Committees shall review individual qualifications and certification and shall address and recommend to the certifying official decertification for anyone they have reviewed for certification. If the review occurs at the forest level, the individual reviewed shall have appeal rights with the Regional Qualification Review Committee. If the review occurs at the regional level, the individual reviewed shall have appeal rights with the Washington Office Qualification Review Committee.

5. <u>Individuals Relieved from Fire Assignment</u>. Individuals who have been relieved from an assignment shall not be reassigned to any incident until the certifying official approves the suitability of the individual to perform the duties associated with the qualifications for the position.

6. <u>Interagency Teams</u>. Interagency teams or groups fall outside Forest Service authority. These teams or groups function and operate at the sole discretion of the chartering group. Teams or groups may be formed, disbanded, held in abeyance, or re-formed at the discretion of the appropriate level of the chartering interagency group, according to applicable standards for each team. Examples of these interagency teams or groups are:

a. Area Command teams chartered and formed by the National Multi-Agency Coordinating Group.

b. National Type 1 teams chartered by geographic area coordinating groups.

    c. Area Type 2 teams chartered by geographic area coordinating groups or by an individual sub-geographic area group.

  7. Crews. Type 1 crews are decertified according to procedures developed nationally. Type 2 crews are decertified on a geographic or sub-geographic area basis.

2.3 - PHYSICAL FITNESS STANDARDS AND DEFINITIONS

Minimum physical fitness standards for positions are contained within the PMS 310-1. Fitness standards for technical specialist positions can be found in Chapter 2, pt 2.

2.31 - Physical Fitness Requirements

Requirements for physical fitness are identified as arduous, moderate, light, and none required. Reference the Wildland Fire Qualification System Guide (310-1) for descriptions.

Personnel taking the work capacity test (WCT) will only complete the level of testing (pack, field, walk) required by the highest fitness level identified for a position on their incident qualification card. To further clarify, employees shall not take the WCT unless they have an incident qualification that requires it, and only at the fitness level required by that position as identified in this document.

For any position assigned to the fireline for non-suppression tasks, the required physical fitness level shall be "Light". Visitors to the line are not "assigned to the line for non-suppression tasks" and therefore are subject to incident commander discretion and/or the guidelines as addressed in "Visits to the Fireline", "Non-Escorted Visits" and/or "Escorted Visits" in the

Interagency Standards for Fire and Fire Aviation Operations (Red Book), if applicable.

2.32 - Physical Fitness Measurement

The work capacity test is the physical fitness measurement recognized by the Forest Service.

2.33 - Fitness Development

Fire personnel required to meet the arduous level for their assigned wildland fire positions shall be provided official time for rigorous exercise to prepare for and maintain the arduous level:

1. Fire funded employees (assigned to fire crews and identified in the Fire Management Action Plan) shall be allowed up to 5 hours per week of physical training when not engaged in wildland fire operations.

2. Employees not funded by fire shall be allowed up to 3 hours per week of physical training, consistent with the local wellness programs. The applicable time is up to 6 months prior to scheduled testing and after passing the test, until the end of the proclaimed fire season. Fire program funds may be used for approved physical training time based on guaranteed availability for fire-related assignments.

The employees and their supervisors shall discuss and schedule opportunities during the work week for fitness activities.

2.4 - MEDICAL STANDARDS [RESERVED]

2.5 - HEALTH SCREENING QUESTIONNAIRE [RESERVED]

2.6 - INCIDENT QUALIFICATION CARD

2.61 - PMS 310-3, Incident Qualification Card Signing Authority

Signing Authority for Incident Response Qualifications Cards resides in the FSH 5109.17, Zero Code, 04-Responsibility.

Certification and documentation requirements must be met prior to signature of Form PMS 310-3, Incident Qualification Card.

Qualification and certification of employees shall be performed by the "hosting" Forest which mobilizes the employee. This includes employees in the Information Services Organization (ISO), Albuquerque Service Center (ASC), Research Stations, Law Enforcement and Investigations, and so forth.

For further direction on delegations of authority and responsibility for positions in the Fire and Aviation Management Program, see FSM 1230 and FSM 5100.

Incident Qualifications Card expiration dates:

> a. Positions requiring work capacity tests (WCT) are valid through the fitness expiration date listed on the card.

> b. Positions not requiring WCT for issuance are valid for 12 months from the date the card was signed by a certifying official.

2.7 - POSITION QUALIFICATIONS

2.71 - NWCG Positions

The NWCG Operations and Workforce Development Committee (OWDC) has established minimum position qualification standards for four of the position categories: Incident Command System, Wildland Fire, Incident Support, and Associated Activities. Those minimum standards are contained within the PMS 310-1.

Chapter 2 pt 1, Exhibit 01 provides the minimum position standards contained within the 310-1, as well as the Forest Service supplement including training requirements, experience, physical requirements, and other positions meeting currency requirements for the four position categories.

Chapter 2 pt 2 provides the minimum position standards that the Forest Service has established for the technical specialist positions.

2.71 - Exhibit 01

POSITION QUALIFICATIONS SECTIONS
Command and General Staff Positions

AREA COMMANDER (ACDR)
Position Category: Incident Command System

310-1 REQUIRED TRAINING
S-620 Area Command

FOREST SERVICE ADDITIONAL REQUIRED TRAINING:
IS-700 National Incident Management System (NIMS), An Introduction
IS-800 National Response Plan (NRP), An Introduction

310-1 REQUIRED EXPERIENCE
Satisfactory performance as an Assistant Area Commander Planning (ACPC)

    +

Successful position performance as an Area Commander (ACDR) on a wildfire incident

    OR

Satisfactory performance as an Assistant Area Commander Logistics (ACLC)

    +

Successful position performance as an Area Commander (ACDR) on a wildfire incident

    OR

Satisfactory performance as an Incident Commander Type 1 (ICT1) on a National Type 1 Incident Management Team

    +

Successful position performance as an Area Commander (ACDR) on a wildfire incident

310-1 PHYSICAL FITNESS LEVEL
None Required

310-1 OTHER POSITIONS THAT WILL MAINTAIN CURRENCY
Assistant Area Commander Planning (ACPC)
Assistant Area Commander Logistics (ACLC)
Incident Commander Type 1 (ICT1)

310-1 OTHER TRAINING WHICH SUPPORTS DEVELOPMENT OF KNOWLEDGE AND SKILLS
None

Task Book available at:
http://www.nwcg.gov/pms/taskbook/command/command.htm

27.1 - Exhibit 01--Continued

POSITION QUALIFICATIONS SECTIONS
Command and General Staff Positions

INCIDENT COMMANDER TYPE 1 (ICT1)
Position Category: Incident Command System

310-1 REQUIRED TRAINING
S-520 Advanced Incident Management
RT-130 Annual Fireline Safety Refresher

FOREST SERVICE ADDITIONAL REQUIRED TRAINING
IS-700 National Incident Management System (NIMS), An Introduction
IS-800 National Response Plan (NRP), An Introduction

310-1 REQUIRED EXPERIENCE
Satisfactory performance as an Incident Commander Type 2 (ICT2)
        +
Successful position performance as an Incident Commander Type 1 (ICT1) on a wildfire incident

310-1 PHYSICAL FITNESS LEVEL
None Required

310-1 OTHER POSITION ASSIGNMENTS THAT WILL MAINTAIN CURRENCY
Finance/Administration Section Chief Type 1 (FSC1)
Logistics Section Chief Type 1(LSC1)
Operations Section Chief Type 1 (OSC1)
Planning Section Chief Type 1 (PSC1)
Incident Commander Type 2 (ICT2)
Assistant Area Commander Planning (ACPC)
Assistant Area Commander Logistics (ACLC)
Any higher position for which this position is a prerequisite

310-1 OTHER TRAINING WHICH SUPPORTS DEVELOPMENT OF KNOWLEDGE AND SKILLS
None

Task Book available at:
http://www.nwcg.gov/pms/taskbook/command/command.htm

27.1 - Exhibit 01--Continued

POSITION QUALIFICATIONS SECTIONS
Command and General Staff Positions

INCIDENT COMMANDER TYPE 2 (ICT2)
(Position Category: Incident Command System)

310-1 REQUIRED TRAINING
S-420 Command and General Staff
RT-130 Annual Fireline Safety Refresher

FOREST SERVICE ADDITIONAL REQUIRED TRAINING:
IS-700 National Incident Management System (NIMS), An Introduction
IS-800 National Response Plan (NRP), An Introduction

310-1 REQUIRED EXPERIENCE
Satisfactory performance as an Incident Commander Type 3 (ICT3)
+
Satisfactory performance as an Operations Section Chief Type 2 (OSC2)
+
Successful position performance as an Incident Commander Type 2 (ICT2) on a wildfire incident
OR
Satisfactory performance as an Incident Commander Type 3 (ICT3)
+
Satisfactory performance as a Planning Section Chief Type 2 (PSC2)
+
Successful position performance as an Incident Commander Type 2 (ICT2) on a wildfire incident
OR
Satisfactory performance as an Incident Commander Type 3 (ICT3)
+
Satisfactory performance as a Logistics Section Chief Type 2 (LSC2)
+
Successful position performance as an Incident Commander Type 2 (ICT2) on a wildfire incident
OR
Satisfactory performance as an Incident Commander Type 3 (ICT3)
+
Satisfactory performance as a Finance/Administration Section Chief Type 2 (FSC2)
+
Successful position performance as an Incident Commander Type 2 (ICT2) on a wildfire incident

310-1 PHYSICAL FITNESS LEVEL
None Required

27.1 - Exhibit 01--Continued

POSITION QUALIFICATIONS SECTIONS
Command and General Staff Positions

INCIDENT COMMANDER TYPE 2 (ICT2) - CONTINUED
(Position Category: Incident Command System)

310-1 OTHER POSITION ASSIGNMENTS THAT WILL MAINTAIN CURRENCY
Logistics Section Chief Type 2 (LSC2)
Planning Section Chief Type 2 (PSC2)
Finance/Administration Section Chief Type 2 (FSC2)
Operations Section Chief Type 2 (OSC2)
Any higher position for which this position is a prerequisite

310-1 OTHER TRAINING WHICH SUPPORTS DEVELOPMENT OF KNOWLEDGE AND SKILLS
I-400 Advanced Incident Command System
S-400 Incident Commander
L-480 Organizational Leadership in the Fire Service

Task Book available at:
http://www.nwcg.gov/pms/taskbook/command/command.htm

27.1 - Exhibit 01--Continued

POSITION QUALIFICATIONS SECTIONS
Command and General Staff Positions

INCIDENT COMMANDER TYPE 3 (ICT3)
Position Category: Incident Command System

310-1 REQUIRED TRAINING
S-300 Incident Commander Extended Attack
S-390 Introduction to Wildland Fire Behavior Calculations
RT-130 Annual Fireline Safety Refresher

FOREST SERVICE ADDITIONAL REQUIRED TRAINING
IS-700 National Incident Management System (NIMS), An Introduction
IS-800 National Response Plan (NRP), An Introduction
L-381 Incident Leadership

FOREST SERVICE CERTIFICATION
Time Pressure Simulation Assessment (TPSA)

Certifying officials shall require trainee Incident Commander, Type 3's to experience at least one trainee assignment, prior to attending the simulation, in order to prepare the trainee for success and a high standard of performance during the simulation.

FOREST SERVICE ANNUAL REVIEW CONSIDERATIONS
Any ICT3 that is determined by their Certifying Official or Forest Qualification Review Committee (FQRC) as:
1. "Inactive" or lacking quality assignments may be directed to successfully complete a TPSA.
2. Experiencing performance problems may be directed to successfully complete a TPSA.

27.1 - Exhibit 01--Continued

POSITION QUALIFICATIONS SECTIONS
Command and General Staff Positions

INCIDENT COMMANDER TYPE 3 (ICT3) - CONTINUED
Position Category: Incident Command System

310-1 REQUIRED EXPERIENCE
Satisfactory performance as an Incident Commander Type 4 (ICT4)
        +
Satisfactory performance as a Task Force Leader (TFLD)
        +
Successful position performance as an Incident Commander Type 3 (ICT3) on a wildfire incident
        OR
Satisfactory position performance as an Incident Commander Type 4 (ICT4)
        +
Satisfactory performance as any Strike Team Leader (STCR, STEN, STEQ, STEQ)
        +
Satisfactory performance in any two single resource boss positions (one must be CRWB or ENGB)
        +
Successful position performance as an Incident Commander Type 3 (ICT3) on a wildfire incident

310-1 PHYSICAL FITNESS LEVEL
Arduous

310-1 OTHER POSITION ASSIGNMENTS THAT WILL MAINTAIN CURRENCY
Division/Group Supervisor (DIVS)
Task Force Leader (TFLD)
Any Strike Team Leader (STCR, STEN, STEQ, STEQ)
Prescribed Fire Burn Boss Type 1 (RXB1)
Any higher position for which this position is a prerequisite

310-1 OTHER TRAINING WHICH SUPPORTS DEVELOPMENT OF KNOWLEDGE AND SKILLS
L-381 Incident Leadership

Task Book available at:
http://www.nwcg.gov/pms/taskbook/command/command.htm

27.1 - Exhibit 01--Continued

POSITION QUALIFICATIONS SECTIONS
Command and General Staff Positions

INCIDENT COMMANDER TYPE 4 (ICT4)
Position Category: Incident Command System

310-1 REQUIRED TRAINING
S-200 Initial Attack Incident Commander
RT-130 Annual Fireline Safety Refresher

FOREST SERVICE ADDITIONAL REQUIRED TRAINING
IS-700 National Incident Management System (NIMS), An Introduction
L-280 Followership to Leadership
S-215 Fire Operations in the Urban Interface

310-1 REQUIRED EXPERIENCE
Satisfactory performance in one of the following single resource boss positions (CRWB, HEQB, ENGB, FELB, FIRB, HMGB, TRPB)
        +
Successful position performance as an Incident Commander Type 4 (ICT4) on a wildfire incident

310-1 PHYSICAL FITNESS LEVEL
Arduous

310-1 OTHER POSITION ASSIGNMENTS THAT WILL MAINTAIN CURRENCY
Any of the following single resource boss positions (ENGB, CRWB, , HEQB, FELB, FIRB, HMGB, TRPB)
Prescribed Fire Burn Boss Type 2 (RXB2)
Any higher position for which this position is a prerequisite

310-1 OTHER TRAINING WHICH SUPPORTS DEVELOPMENT OF KNOWLEDGE AND SKILLS
S-234 Ignition Operations
S-215 Fire Operations in the Wildland/Urban Interface

Task Book available at:
http://www.nwcg.gov/pms/taskbook/command/command.htm

27.1 - Exhibit 01--Continued

POSITION QUALIFICATIONS SECTIONS
Command and General Staff Positions

INCIDENT COMMANDER TYPE 5 (ICT5)
Position Category: Incident Command System

The ICT5 position is not prerequisite to the NWCG Incident Commander Type 4 (ICT4) position

310-1 REQUIRED TRAINING
S-131 Firefighter Type 1
S-133 Look up, Look Down, Look Around
RT-130 Annual Fireline Safety Refresher

FOREST SERVICE ADDITIONAL REQUIRED TRAINING
IS-700 National Incident Management System (NIMS), An Introduction

310-1 REQUIRED EXPERIENCE
Satisfactory performance as a Firefighter Type 2 (FFT2)

+

Successful position performance as an Incident Commander Type 5 (ICT5) on a wildfire incident.

310-1 PHYSICAL FITNESS LEVEL
Arduous

310-1 OTHER POSITION ASSIGNMENTS THAT WILL MAINTAIN CURRENCY
Incident Commander Type 4 (ICT4)
Firefighter Type 1 (FFT1)

310-1 OTHER TRAINING WHICH SUPPORTS DEVELOPMENT OF KNOWLEDGE AND SKILLS
S-212 Wildland Fire Chain Saws
S-211 Portable Pumps and Water Use

Task Book available at:
http://www.nwcg.gov/pms/taskbook/command/command.htm
Note: the Firefighter Type 1 (FFT1) and Incident Commander Type 5 (ICT5) Position Task Books have been combined. However, the positions have not been combined. The FFT1 tasks are completed only once; additional tasks must be completed to meet the ICT5 level. The FFT1 and ICT5 tasks can be completed simultaneously. The Required Experience is satisfactory performance as a Firefighter Type 2 (FFT2).

27.1 - Exhibit 01--Continued

POSITION QUALIFICATIONS SECTIONS
Command and General Staff Positions

SAFETY OFFICER TYPE 1 (SOF1)
Position Category: Incident Command System

310-1 REQUIRED TRAINING
S-520 Advanced Incident Management
RT-130 Annual Fireline Safety Refresher

FOREST SERVICE ADDITIONAL REQUIRED TRAINING
IS-700 National Incident Management System (NIMS), An Introduction
IS-800 National Response Plan (NRP), An Introduction

310-1 REQUIRED EXPERIENCE
Satisfactory performance as a Safety Officer Type 2 (SOF2)
        +
Successful position performance as a Safety Officer Type 1 (SOF1) on a wildland fire incident

310-1 PHYSICAL FITNESS LEVEL
Moderate

310-1 OTHER POSITION ASSIGNMENTS THAT WILL MAINTAIN CURRENCY
Operations Section Chief Type 2 (OSC2)
Safety Officer Type 2 (SOF2)

310-1 OTHER TRAINING WHICH SUPPORTS DEVELOPMENT OF KNOWLEDGE AND SKILLS
None

Task Book available at:
http://www.nwcg.gov/pms/taskbook/command/command.htm

27.1 - Exhibit 01--Continued

POSITION QUALIFICATIONS SECTIONS
Command and General Staff Positions

SAFETY OFFICER TYPE 2 (SOF2)
Position Category: Incident Command System

310-1 REQUIRED TRAINING
S-420 Command and General Staff
RT-130 Annual Fireline Safety Refresher

FOREST SERVICE ADDITIONAL REQUIRED TRAINING
IS-700 National Incident Management System (NIMS), An Introduction
IS-800 National Response Plan (NRP), An Introduction

310-1 REQUIRED EXPERIENCE
Satisfactory performance as a Division/Group Supervisor (DIVS)
> +

Successful position performance as a Safety Officer Type 2 (SOF2) on a wildland fire incident

310-1 PHYSICAL FITNESS LEVEL
Moderate

310-1 OTHER POSITION ASSIGNMENTS THAT WILL MAINTAIN CURRENCY
Division/Group Supervisor (DIVS)
Safety Officer, Line (SOFR)
Any higher position for which this position is a prerequisite

310-1 OTHER TRAINING WHICH SUPPORTS DEVELOPMENT OF KNOWLEDGE AND SKILLS
I-400 Advanced Incident Command System
S-404 Safety Officer
L-480 Organizational Leadership in the Fire Service

Task Book available at:
http://www.nwcg.gov/pms/taskbook/command/command.htm

POSITION QUALIFICATIONS SECTIONS
Command and General Staff Positions

SAFETY OFFICER LINE (SOFR)
Position Category: Incident Command System
This position is not prerequisite to the NWCG Safety Officer Type 2 (SOF2) position.

310-1 REQUIRED TRAINING
Annual Fireline Safety Refresher (RT-130)

FOREST SERVICE ADDITIONAL REQUIRED TRAINING:
IS-700 National Incident Management System (NIMS), An Introduction

310-1 REQUIRED EXPERIENCE
Satisfactory performance as any Strike Team Leader (STCR, STEN, STEQ, STEQ)
 +
Successful position performance as a Safety Officer, Line (SOFR) on a wildland fire incident
 OR
Satisfactory performance as an Incident Commander Type 4 (ICT4)
 +
Successful position performance as a Safety Officer, Line (SOFR) on a wildland fire incident

310-1 PHYSICAL FITNESS LEVEL
Moderate

310-1 OTHER POSITION ASSIGNMENTS THAT WILL MAINTAIN CURRENCY
Safety Officer Type 2 (SOF2)
Incident Commander Type 4 (ICT4)
Task Force Leader (TFLD)
Any Strike Team Leader (STCR, STEN, STEQ, STEQ)

310-1 OTHER TRAINING WHICH SUPPORTS DEVELOPMENT OF KNOWLEDGE AND SKILLS
I-300 Intermediate Incident Command System

Task Book available at:
http://www.nwcg.gov/pms/taskbook/command/command.htm

27.1 - Exhibit 01--Continued

POSITION QUALIFICATIONS SECTIONS
Command and General Staff Positions

LIAISON OFFICER (LOFR)*
Position Category: Incident Command System

310-1 REQUIRED TRAINING
None

FOREST SERVICE ADDITIONAL REQUIRED TRAINING:
I-100 Introduction to Incident Command System**
I-200 Basic Incident Command System
I-300 Intermediate Incident Command System
I-400 Advanced Incident Command System
IS-700 National Incident Management System (NIMS), An Introduction
IS-800 National Response Plan (NRP), An Introduction

310-1 REQUIRED EXPERIENCE
Successful position performance as a Liaison Officer (LOFR)

310-1 PHYSICAL FITNESS LEVEL
None Required

310-1 OTHER POSITION ASSIGNMENTS THAT WILL MAINTAIN CURRENCY
Agency Representative (AREP)

310-1 OTHER TRAINING WHICH SUPPORTS DEVELOPMENT OF KNOWLEDGE AND SKILLS
I-400 Advanced Incident Command System
L-180 Human Factors on the Fireline

* Designates agency established position qualifications
**I-100 Online link: http://training.nwcg.gov/classes/i100.htm
Task Book available at:
http://www.nwcg.gov/pms/taskbook/command/command.htm

27.1 - Exhibit 01--Continued

POSITION QUALIFICATIONS SECTIONS
Command and General Staff Positions

PUBLIC INFORMATION OFFICER TYPE 1 (PIO1)
Position Category: Incident Command System

310-1 REQUIRED TRAINING
S-520 Advanced Incident Management (S-520)
RT-130 Annual Wildland Fire Safety Refresher (required annually after initial training)

FOREST SERVICE ADDITIONAL REQUIRED TRAINING
IS-700 National Incident Management System (NIMS), An Introduction
IS-800 National Response Plan (NRP), An Introduction

310-1 REQUIRED EXPERIENCE
Satisfactory performance as a Public Information Officer Type 2 (PIO2)
+
Successful position performance as a Public Information Officer Type 1 (PIO1)

310-1 PHYSICAL FITNESS LEVEL
None Required

310-1 OTHER POSITION ASSIGNMENTS THAT WILL MAINTAIN CURRENCY
Public Information Officer Type 2 (PIO2)

310-1 OTHER TRAINING WHICH SUPPORTS DEVELOPMENT OF KNOWLEDGE AND SKILLS
None

Task Book available at:
http://www.nwcg.gov/pms/taskbook/command/command.htm

27.1 - Exhibit 01--Continued

POSITION QUALIFICATIONS SECTIONS
Command and General Staff Positions

PUBLIC INFORMATION OFFICER TYPE 2 (PIO2)
Position Category: Incident Command System

310-1 REQUIRED TRAINING
S-420 Command and General Staff (S-420)
S-190 Introduction to Wildland Fire Behavior (S-190)
RT-130 Annual Wildland Fire Safety Refresher Training (required annually after initial training)

FOREST SERVICE ADDITIONAL REQUIRED TRAINING
I-100 Introduction to Incident Command System*
I-200 Basic Incident Command System
I-300 Intermediate Incident Command System
I-400 Advanced Incident Command System
IS-700 National Incident Management System (NIMS), An Introduction
IS-800 National Response Plan (NRP), An Introduction

310-1 REQUIRED EXPERIENCE
Successful position performance as a Public Information Officer (PIOF)
       +
Successful position performance as a Public Information Officer Type 2 (PIO2)

310-1 PHYSICAL FITNESS LEVEL
None Required

310-1 OTHER POSITION ASSIGNMENTS THAT WILL MAINTAIN CURRENCY
Public Information Officer (PIOF)
Any higher position for which this position is a prerequisite

310-1 OTHER TRAINING WHICH SUPPORTS DEVELOPMENT OF KNOWLEDGE AND SKILLS
I-400 Advanced Incident Command System
S-403 Information Officer

Task Book available at:
http://www.nwcg.gov/pms/taskbook/command/command.htm
*I-100 Online link: *http://training.nwcg.gov/classes/i100.htm*

27.1 - Exhibit 01--Continued

POSITION QUALIFICATIONS SECTIONS
Command and General Staff Positions

PUBLIC INFORMATION OFFICER (PIOF)
Position Category: Incident Command System

This position **IS NOW A** *prerequisite to the NWCG Public Information Officer Type 2 (PIO2) position*

310-1 REQUIRED TRAINING
I-100 Introduction to Incident Command System*
L-180 Human Factors on the Fireline
RT-130 Annual Fireline Refresher Training (required annually after initial training)
S-130 Firefighter Training
S-190 Introduction to Wildland Fire Behavior
S-203 Introduction to Incident Information

FOREST SERVICE ADDITIONAL REQUIRED TRAINING
IS-700 National Incident Management System (NIMS), An Introduction

310-1 REQUIRED EXPERIENCE
Successful position performance as a Public Information Officer (PIOF)

310-1 PHYSICAL FITNESS LEVEL
None Required

310-1 OTHER POSITION ASSIGNMENTS THAT WILL MAINTAIN CURRENCY
Public Information Officer Type 2 (PIO2)

310-1 OTHER TRAINING WHICH SUPPORTS DEVELOPMENT OF KNOWLEDGE AND SKILLS
I-300 Intermediate Incident Command System
S-110 Basic Wildland Fire Orientation

Task Book available at:
http://www.nwcg.gov/pms/taskbook/command/command.htm
*I-100 Online link: *http://training.nwcg.gov/classes/i100.htm*

27.1 - Exhibit 01--Continued

POSITION QUALIFICATIONS SECTIONS
Command and General Staff Positions

AGENCY REPRESENTATIVE (AREP)*
Position Category: Incident Command System

310-1 REQUIRED TRAINING
None

FOREST SERVICE ADDITIONAL REQUIRED TRAINING
I-100 Introduction to Incident Command System**
I-200 Basic Incident Command System
IS-700 National Incident Management System (NIMS), An Introduction
IS-800 National Response Plan (NRP), An Introduction

310-1 REQUIRED EXPERIENCE
Agency established

FOREST SERVICE DESIRED EXPERIENCE:
Desirable skills include a thorough orientation to applicable cooperative agreements and knowledge of Forest Service policies and procedures

310-1 PHYSICAL FITNESS LEVEL
None Required

310-1 OTHER POSITION ASSIGNMENTS THAT WILL MAINTAIN CURRENCY
Liaison Officer (LOFR)

310-1 OTHER TRAINING WHICH SUPPORTS DEVELOPMENT OF KNOWLEDGE AND SKILLS
L-180 Human Factors on the Fireline (L-180)

* Designates agency established position qualification
**I-100 Online link: *http://training.nwcg.gov/classes/i100.htm*

27.1 - Exhibit 01--Continued

POSITION QUALIFICATIONS SECTIONS
Command and General Staff Positions

INTERAGENCY RESOURCE REPRESENTATIVE (IARR)*
Position Category: Incident Command System

310-1 REQUIRED TRAINING
None

FOREST SERVICE ADDITIONAL REQUIRED TRAINING
IS-700 National Incident Management System (NIMS), An Introduction

310-1 REQUIRED EXPERIENCE
Agency established

FOREST SERVICE REQUIRED EXPERIENCE
Satisfactory position performance as a Single Resource Boss, Crew (CRWB) or Engine (ENGB)
    AND
 Successful position performance as an Interagency Resource Representative (IARR)

310-1 PHYSICAL FITNESS LEVEL
None Required

310-1 OTHER POSITION ASSIGNMENTS THAT WILL MAINTAIN CURRENCY
Crew Representative (CREP)

310-1 OTHER TRAINING WHICH SUPPORTS DEVELOPMENT OF KNOWLEDGE AND SKILLS
L-180 Human Factors on the Fireline

* Designates agency established position qualifications

27.1 - Exhibit 01--Continued

POSITION QUALIFICATION SECTIONS
Command and General Staff Positions

PRESCRIBED FIRE MANAGER TYPE 1 (RXM1)
Position Category: Wildland Fire

310-1 REQUIRED TRAINING
RT-130 Annual Fireline Safety Refresher

FOREST SERVICE ADDITIONAL REQUIRED TRAINING
IS-700 National Incident Management System (NIMS), An Introduction

310-1 REQUIRED EXPERIENCE
Satisfactory performance as a Prescribed Fire Burn Boss Type 1 (RXB1)
+
Successful position performance as a Prescribed Fire Manager Type 1 (RXM1) on a Prescribed Fire Incident

310-1 PHYSICAL FITNESS LEVEL
None Required

310-1 OTHER POSITION ASSIGNMENTS THAT WILL MAINTAIN CURRENCY
Prescribed Fire Manager Type 2 (RXM2)

310-1 OTHER TRAINING WHICH SUPPORTS DEVELOPMENT OF KNOWLEDGE AND SKILLS
I-300 Intermediate Incident Command System
M-581 Fire Program Management

Task Book available at:
http://www.nwcg.gov/pms/taskbook/command/command.htm

27.1 - Exhibit 01--Continued

POSITION QUALIFICATIONS SECTIONS
Command and General Staff Positions

PRESCRIBED FIRE MANAGER TYPE 2 (RXM2)
Position Category: Wildland Fire

310-1 REQUIRED TRAINING
RT-130 Annual Fireline Safety Refresher

FOREST SERVICE ADDITIONAL REQUIRED TRAINING
IS-700 National Incident Management System (NIMS), An Introduction

310-1 REQUIRED EXPERIENCE
Satisfactory performance as a Prescribed Fire Burn Boss Type 2 (RXB2)
+
Successful position performance as a Prescribed Fire Manager Type 2 (RXM2) on a
Prescribed Fire Incident

310-1 PHYSICAL FITNESS LEVEL
None Required

310-1 OTHER POSITION ASSIGNMENTS THAT WILL MAINTAIN CURRENCY
Prescribed Fire Burn Boss Type 2 (RXB2)

310-1 OTHER TRAINING WHICH SUPPORTS DEVELOPMENT OF KNOWLEDGE AND SKILLS
I-300 Intermediate Incident Command System
M-581 Fire Program Management

Task Book available at:
http://www.nwcg.gov/pms/taskbook/command/command.htm

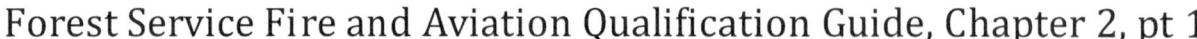

Note: SOPL has been moved to the Planning Positions Section

27.1 - Exhibit 01--Continued

POSITION QUALIFICATIONS SECTIONS
Command and General Staff Positions

PRESCRIBED FIRE BURN BOSS TYPE 1 (RXB1)
Position Category: Wildland Fire

310-1 REQUIRED TRAINING
S-490 Advanced Wildland Fire Behavior Calculations
RT-130 Annual Fireline Safety Refresher

FOREST SERVICE ADDITIONAL REQUIRED TRAINING
IS-700 National Incident Management System (NIMS), An Introduction
RX-410 Smoke Management Techniques

FOREST SERVICE CERTIFICATION
RT-300FS – Prescribed Fire Workshop (Biennial)

310-1 REQUIRED EXPERIENCE
Satisfactory performance as a Prescribed Fire Burn Boss Type 2 (RXB2)
> +
Successful position performance as a Prescribed Fire Burn Boss Type 1 (RXB1) on a
prescribed fire incident

310-1 PHYSICAL FITNESS LEVEL
Light

310-1 OTHER POSITION ASSIGNMENTS THAT WILL MAINTAIN CURRENCY
Prescribed Fire Burn Boss Type 2 (RXB2)
Any higher position for which this position is a prerequisite

310-1 OTHER TRAINING WHICH SUPPORTS DEVELOPMENT OF KNOWLEDGE AND SKILLS
M-581 Fire Program Management
M-580 Fire in Ecosystem Management
RX-510 Advanced Fire Effects
RX-410 Smoke Management Techniques

Task Book available at:
http://www.nwcg.gov/pms/taskbook/command/command.htm

27.1 - Exhibit 01--Continued

POSITION QUALIFICATION SECTIONS
Command and General Staff Positions

PRESCRIBED FIRE BURN BOSS TYPE 2 (RXB2)
Position Category: Wildland Fire

310-1 REQUIRED TRAINING
S-390 Introduction to Wildland Fire Behavior Calculations
RT-130 Annual Fireline Safety Refresher

FOREST SERVICE ADDITIONAL REQUIRED TRAINING
IS-700 National Incident Management System (NIMS), An Introduction
RX-301 Prescribed Fire Implementation
RX-310 Introduction to Fire Effects
RX-341 Prescribed Fire Burn Plan Preparation
RX-410 Smoke Management Techniques

FOREST SERVICE CERTIFICATION
RT-300FS – Prescribed Fire Workshop (Biennial)

310-1 REQUIRED EXPERIENCE
Satisfactory performance as a Firing Boss, Single Resource (FIRB)
        +
Satisfactory performance as an Incident Commander Type 4 (ICT4)
        +
Successful position performance as a Prescribed Fire Burn Boss Type 2 (RXB2) on a prescribed fire incident

310-1 PHYSICAL FITNESS LEVEL
Moderate

310-1 OTHER POSITION ASSIGNMENTS THAT WILL MAINTAIN CURRENCY
Any higher position for which this position is a prerequisite

310-1 OTHER TRAINING WHICH SUPPORTS DEVELOPMENT OF KNOWLEDGE AND SKILLS
L-380 Fireline Leadership

Task Book available at:
http://www.nwcg.gov/pms/taskbook/command/command.htm

27.1 - Exhibit 01--Continued

POSITION QUALIFICATIONS SECTIONS
Operations Positions

OPERATIONS SECTION CHIEF TYPE 1 (OSC1)
Position Category: Incident Command System

310-1 REQUIRED TRAINING
S-520 Advanced Incident Management
RT-130 Annual Fireline Safety Refresher

FOREST SERVICE ADDITIONAL REQUIRED TRAINING
IS-700 National Incident Management System (NIMS), An Introduction
IS-800 National Response Plan (NRP), An Introduction

310-1 REQUIRED EXPERIENCE
Satisfactory performance as an Operations Section Chief Type 2 (OSC2)
          +
Successful position performance as an Operations Section Chief Type 1 (OSC1) on a wildfire incident

310-1 PHYSICAL FITNESS LEVEL
Moderate

310-1 OTHER POSITION ASSIGNMENTS THAT WILL MAINTAIN CURRENCY
Operations Section Chief Type 2 (OSC2)
Operations Branch Director (OPBD)
Incident Commander Type 1 (ICT1)
Any higher position for which this position is a prerequisite

310-1 OTHER TRAINING WHICH SUPPORTS DEVELOPMENT OF KNOWLEDGE AND SKILLS
None

Task Book available at:
http://www.nwcg.gov/pms/taskbook/operations/operation.htm

27.1 - Exhibit 01--Continued

POSITION QUALIFICATIONS SECTIONS
Operations Positions

OPERATIONS SECTION CHIEF TYPE 2 (OSC2)
Position Category: Incident Command System

310-1 REQUIRED TRAINING
S-420 Command and General Staff
RT-130 Annual Fireline Safety Refresher

FOREST SERVICE ADDITIONAL REQUIRED TRAINING
IS-700 National Incident Management System (NIMS), An Introduction
IS-800 National Response Plan (NRP), An Introduction
S-430 Operations Section Chief

310-1 REQUIRED EXPERIENCE
Satisfactory performance as a Division/Group Supervisor (DIVS)

+

Successful position performance as an Operations Section Chief Type 2 (OSC2) on a wildfire incident

310-1 PHYSICAL FITNESS LEVEL
Moderate

310-1 OTHER POSITION ASSIGNMENTS THAT WILL MAINTAIN CURRENCY
Division/Group Supervisor (DIVS)
Incident Commander Type 3 (ICT3)
Any higher position for which this position is a prerequisite

310-1 OTHER TRAINING WHICH SUPPORTS DEVELOPMENT OF KNOWLEDGE AND SKILLS
I-400 Advanced Incident Command System
S-430 Operations Section Chief
L-480 Organizational Leadership in the Fire Service

Task Book available at:
http://www.nwcg.gov/pms/taskbook/operations/operation.htm

POSITION QUALIFICATIONS SECTIONS
Operations Positions

OPERATIONS BRANCH DIRECTOR (OPBD)
Position Category: Incident Command System

310-1 REQUIRED TRAINING
RT-130 Annual Fireline Safety Refresher

FOREST SERVICE ADDITIONAL REQUIRED TRAINING
IS-700 National Incident Management System (NIMS), An Introduction
IS-800 National Response Plan (NRP), An Introduction

310-1 REQUIRED EXPERIENCE
Satisfactory performance as an Operations Section Chief Type 2 (OSC2)

310-1 PHYSICAL FITNESS LEVEL
Moderate

310-1 OTHER POSITION ASSIGNMENTS THAT WILL MAINTAIN CURRENCY
Division/Group Supervisor (DIVS)
Incident Commander Type 3 (ICT3)
Any higher position for which this position is a prerequisite

310-1 OTHER TRAINING WHICH SUPPORTS DEVELOPMENT OF KNOWLEDGE AND SKILLS
None

27.1 - Exhibit 01--Continued

POSITION QUALIFICATIONS SECTIONS
Operations Positions

STRUCTURE PROTECTION SPECIALIST (STPS)
Position Category: Wildland Fire

310-1 REQUIRED TRAINING
RT-130 Annual Fireline Safety Refresher

FOREST SERVICE ADDITIONAL REQUIRED TRAINING
IS-700 National Incident Management System (NIMS), An Introduction

310-1 REQUIRED EXPERIENCE
Satisfactory performance as a Division/Group Supervisor (DIVS)

+

Successful position performance as a Structure Protection Specialist (STPS) on a wildland fire incident

OR

Satisfactory performance as an Incident Commander Type 3 (ICT3)

+

Successful position performance as a Structure Protection Specialist (STPS) on a wildland fire incident

310-1 PHYSICAL FITNESS LEVEL
Moderate

310-1 OTHER POSITION ASSIGNMENTS THAT WILL MAINTAIN CURRENCY
Operations Branch Director (OPBD)
Operations Section Chief Type 2 (OSC2)
Division/Group Supervisor (DIVS)
Incident Commander Type 3 (ICT3)

310-1 OTHER TRAINING WHICH SUPPORTS DEVELOPMENT OF KNOWLEDGE AND SKILLS
None

Task Book available at:
http://www.nwcg.gov/pms/taskbook/operations/operation.htm

POSITION QUALIFICATIONS SECTIONS
Operations Positions

DIVISION/GROUP SUPERVISOR (DIVS)
Position Category: Incident Command System

310-1 REQUIRED TRAINING
Introduction to Wildland Fire Behavior Calculations (S-390)
Division/Group Supervisor (S-339)
Annual Fireline Safety Refresher (RT-130)

FOREST SERVICE ADDITIONAL REQUIRED TRAINING
L-381 Incident Leadership
IS-700 National Incident Management System (NIMS), An Introduction
IS-800 National Response Plan (NRP), An Introduction

310-1 REQUIRED EXPERIENCE
Satisfactory performance as a Task Force Leader (TFLD)

    +

Successful position performance as a Division/Group Supervisor (DIVS) on a wildland fire incident

    OR

Satisfactory performance as an Incident Commander Type 3 (ICT3)

    +

Successful position performance as a Division/Group Supervisor (DIVS) on a wildland fire incident

    OR

Satisfactory performance as an Incident Commander Type 4 (ICT4)

    +

Satisfactory performance in any two Strike Team Leader positions (one must be STCR or STEN)

    +

Successful position performance as a Division/Group Supervisor (DIVS) on a wildland fire incident

310-1 PHYSICAL FITNESS LEVEL
Arduous

27.1 - Exhibit 01--Continued

POSITION QUALIFICATIONS SECTIONS
Operations Positions

DIVISION/GROUP SUPERVISOR (DIVS) - CONTINUED
Position Category: Incident Command System

310-1 OTHER POSITION ASSIGNMENTS THAT WILL MAINTAIN CURRENCY
Task Force Leader (TFLD)
Strike Team Leader Crew (STCR) or Strike Team Leader Engine (STEN)
Incident Commander Type 3 (ICT3)
Any higher position for which this position is a prerequisite

310-1 OTHER TRAINING WHICH SUPPORTS DEVELOPMENT OF KNOWLEDGE AND SKILLS
L-381 Incident Leadership

Task Book available at:
http://www.nwcg.gov/pms/taskbook/operations/operation.htm

27.1 - Exhibit 01--Continued

POSITION QUALIFICATIONS SECTIONS
Operations Positions

TASK FORCE LEADER (TFLD)
Position Category: Incident Command System

310-1 REQUIRED TRAINING
S-330 Task Force/Strike Team Leader
S-215 Fire Operations in the Wildland/Urban Interface
RT-130 Annual Fireline Safety Refresher

FOREST SERVICE ADDITIONAL TRAINING
IS-700 National Incident Management System (NIMS), An Introduction
IS-800 National Response Plan (NRP), An Introduction
L-380 Fireline Leadership*
S-336 Fire Suppression Tactics or equivalent training (if prerequisite experience has been attained by * below)
S-390 Introduction to Wildland Fire Behavior Calculations
I-300 Intermediate Incident Command System*

* If required experience has been attained through completion of two single resource boss (one of which must be crew or engine)

310-1 REQUIRED EXPERIENCE
Satisfactory performance as any Strike Team Leader (STCR, STEN, STEQ, STEQ)
        +
Successful position performance as a Task Force Leader (TFLD) on a wildland fire incident
        OR
*Satisfactory performance in any two single resource boss positions (one must be CRWB or ENGB)
        +
Satisfactory performance as an Incident Commander Type 4 (ICT4)
        +
Successful position performance as a Task Force Leader (TFLD) on a wildland fire incident

310-1 PHYSICAL FITNESS LEVEL
Arduous

27.1 - Exhibit 01--Continued

POSITION QUALIFICATIONS SECTIONS
Operations Positions

TASK FORCE LEADER (TFLD) - CONTINUED
Position Category: Incident Command System

310-1 OTHER POSITION ASSIGNMENTS THAT WILL MAINTAIN CURRENCY
Incident Commander Type 4 (ICT4)
Any Strike Team Leader (STCR, STEN, STEQ, STEQ)
Any higher position for which this position is a prerequisite

310-1 OTHER TRAINING WHICH SUPPORTS DEVELOPMENT OF KNOWLEDGE AND SKILLS
I-300 Intermediate ICS
L-380 Fireline Leadership
S-336 Tactical Decision Making in Wildland Fire

Task Book available at:
http://www.nwcg.gov/pms/taskbook/operations/operation.htm

Note: The Task Force Leader (TFLD) and Strike Team Leader (STEQ, STEN, STCR, STEQ) Position Task Books have been combined. However, the positions have not been combined. Strike Team Leader tasks are completed only once. The Required Experience for TFLD must be met prior to completing additional TFLD tasks.

POSITION QUALIFICATIONS SECTIONS
Operations Positions

STRIKE TEAM LEADER EQUIPMENT (STEQ)
(formerly known as TRACTOR/PLOW (STPL))
Position Category: Incident Command System

310-1 REQUIRED TRAINING
S-330 Task Force/Strike Team Leader
S-215 Fire Operations in the Wildland/Urban Interface
RT-130 Annual Fireline Safety Refresher

FOREST SERVICE ADDITIONAL REQUIRED TRAINING
I-300 Intermediate Incident Command System
IS-700 National Incident Management System (NIMS), An Introduction
IS-800 National Response Plan (NRP), An Introduction
L-380 Fireline Leadership
S-390 Introduction to Wildland Fire Behavior Calculations

310-1 REQUIRED EXPERIENCE
Satisfactory performance as a Tractor/Plow Boss, Single Resource (TRPB)
        +
Successful position performance as a Strike Team Leader Tractor/Plow (STEQ) on a wildland
fire incident

310-1 PHYSICAL FITNESS LEVEL
Moderate

310-1 OTHER POSITION ASSIGNMENTS THAT WILL MAINTAIN CURRENCY
Strike Team Leader (STCR, STEQ, STEN)
Any higher position for which this position is a prerequisite

**310-1 OTHER TRAINING WHICH SUPPORTS DEVELOPMENT OF KNOWLEDGE
AND SKILLS**
I-300 Intermediate Incident Command System
L-480 Fireline Leadership

27.1 - Exhibit 01--Continued

POSITION QUALIFICATIONS SECTIONS
Operations Positions

STRIKE TEAM LEADER EQUIPMENT (STEQ)
Position Category: Incident Command System

FOREST SERVICE OTHER TRAINING WHICH SUPPORTS DEVELOPMENT OF KNOWLEDGE AND SKILLS
Geographic Area Intermediate Air Operations

Task Book available at:
http://www.nwcg.gov/pms/taskbook/operations/operation.htm

Note: The Task force Leader (TFLD) and Strike Team Leader (STCR, STEQ, STEN, STEQ) Position Task Books have been combined. However, the positions have not been combined. Strike Team Leader tasks are completed only once. The Required Experience for TFLD must be met prior to completing additional TFLD tasks.

Upon satisfactory performance in the prerequisite Single Resource Boss position, the specific strike team leader task book may be initiated. Once qualified as a Strike Team Leader, any additional single resource boss qualifications will also qualify the individual in that corresponding strike team leader position - without having to complete the Strike Team Leader PTB for the new position - once agency certification is documented on the PTB Verification/Certification page.

27.1 - Exhibit 01--Continued

POSITION QUALIFICATIONS SECTIONS
Operations Positions

STRIKE TEAM LEADER EQUIPMENT (STEQ)
(Formerly known as Strike Team Leader Dozer (STDZ))
Position Category: Incident Command System

310-1 REQUIRED TRAINING
S-330 Task Force/Strike Team Leader
S-215 Fire Operations in the Wildland/Urban Interface
RT-130 Annual Fireline Safety Refresher

FOREST SERVICE ADDITIONAL REQUIRED TRAINING
I-300 Intermediate Incident Command System
IS-700 National Incident Management System (NIMS), An Introduction
IS-800 National Response Plan (NRP), An Introduction
L-380 Fireline Leadership
S-336 Fire Suppression Tactics or equivalent training (see chapter 30 of this Handbook)
S-390 Introduction to Wildland Fire Behavior Calculations

310-1 REQUIRED EXPERIENCE
Satisfactory performance as a Dozer Boss, Single Resource (HEQB)
          +
Successful position performance as a Strike Team Leader Dozer (STEQ) on a wildland fire incident

310-1 PHYSICAL FITNESS LEVEL
Moderate

310-1 OTHER POSITION ASSIGNMENTS THAT WILL MAINTAIN CURRENCY
Strike Team Leader (STCR, STEN, STEQ)
Any higher position for which this position is a prerequisite

310-1 OTHER TRAINING WHICH SUPPORTS DEVELOPMENT OF KNOWLEDGE AND SKILLS
I-300 Intermediate Incident Command System
L-380 Fireline Leadership

27.1 - Exhibit 01--Continued

POSITION QUALIFICATIONS SECTIONS
Operations Positions

STRIKE TEAM LEADER EQUIPMENT (STEQ)
Position Category: Incident Command System

FOREST SERVICE OTHER TRAINING WHICH SUPPORTS DEVELOPMENT OF KNOWLEDGE AND SKILLS:
Geographic Area Intermediate Air Operations

Task Book available at:
http://www.nwcg.gov/pms/taskbook/operations/operation.htm

Note: The Task force Leader (TFLD) and Strike Team Leader (STCR, STEQ, STEN, STEQ) Position Task Books have been combined. However, the positions have not been combined. Strike Team Leader tasks are completed only once. The Required Experience for TFLD must be met prior to completing additional TFLD tasks.

Upon satisfactory performance in the prerequisite Single Resource Boss position, the specific strike team leader task book may be initiated. Once qualified as a Strike Team Leader, any additional single resource boss qualifications will also qualify the individual in that corresponding strike team leader position - without having to complete the Strike Team Leader PTB for the new position - once agency certification is documented on the PTB Verification/Certification page.

27.1 - Exhibit 01--Continued

POSITION QUALIFICATIONS SECTIONS
Operations Positions

STRIKE TEAM LEADER ENGINE (STEN)
Position Category: Incident Command System

310-1 REQUIRED TRAINING
S-330 Task Force/Strike Team Leader
S-215 Fire Operations in the Wildland/Urban Interface
RT-130 Annual Fireline Safety Refresher

FOREST SERVICE ADDITIONAL REQUIRED TRAINING
I-300 Intermediate Incident Command System
IS-700 National Incident Management System (NIMS), An Introduction
IS-800 National Response Plan (NRP), An Introduction
L-380 Fireline Leadership
S-336 Fire Suppression Tactics or equivalent training (see chapter 30 of this Handbook)
S-390 Introduction to Wildland Fire Behavior Calculations

310-1 REQUIRED EXPERIENCE
Satisfactory performance as an Engine Boss, Single Resource (ENGB)
        +
Successful position performance as a Strike Team Leader Engine (STEN) on a wildland fire incident

310-1 PHYSICAL FITNESS LEVEL
Moderate

310-1 OTHER POSITION ASSIGNMENTS THAT WILL MAINTAIN CURRENCY
Strike Team Leader (STCR, STEQ, STEQ)
Any higher position for which this position is a prerequisite

OTHER TRAINING WHICH SUPPORTS DEVELOPMENT OF KNOWLEDGE AND SKILLS
I-300 Intermediate Incident Command System
L-380 Fireline Leadership

27.1 - Exhibit 01--Continued

POSITION QUALIFICATIONS SECTIONS
Operations Positions

STRIKE TEAM LEADER ENGINE (STEN) - CONTINUED
Position Category: Incident Command System

FOREST SERVICE OTHER TRAINING WHICH SUPPORTS DEVELOPMENT OF KNOWLEDGE AND SKILLS
Geographic Area Intermediate Air Operations

Task Book available at:
http://www.nwcg.gov/pms/taskbook/operations/operation.htm

Note: The Task force Leader (TFLD) and Strike Team Leader (STCR, STEQ, STEN, STEQ) Position Task Books have been combined. However, the positions have not been combined. Strike Team Leader tasks are completed only once. The Required Experience for TFLD must be met prior to completing additional TFLD tasks.

Upon satisfactory performance in the prerequisite Single Resource Boss position, the specific strike team leader task book may be initiated. Once qualified as a Strike Team Leader, any additional single resource boss qualifications will also qualify the individual in that corresponding strike team leader position - without having to complete the Strike Team Leader PTB for the new position - once agency certification is documented on the PTB Verification/Certification page.

27.1 - Exhibit 01--Continued

POSITION QUALIFICATIONS SECTIONS
Operations Positions

STRIKE TEAM LEADER CREW (STCR)
Position Category: Incident Command System

310-1 REQUIRED TRAINING
S-330 Task Force/Strike Team Leader
S-215 Fire Operations in the Wildland/Urban Interface
RT-130 Annual Fireline Safety Refresher

FOREST SERVICE ADDITIONAL REQUIRED TRAINING
I-300 Intermediate Incident Command System
IS-700 National Incident Management System (NIMS), An Introduction
IS-800 National Response Plan (NRP), An Introduction
L-380 Fireline Leadership
S-336 Fire Suppression Tactics or equivalent training (see chapter 30 of this Handbook)
S-390 Introduction to Wildland Fire Behavior Calculations

310-1 REQUIRED EXPERIENCE
Satisfactory performance as a Crew Boss, Single Resource (CRWB)
        +
Successful position performance as a Strike Team Leader Crew (STCR) on a wildland fire incident

310-1 PHYSICAL FITNESS LEVEL
Arduous

310-1 OTHER POSITION ASSIGNMENTS THAT WILL MAINTAIN CURRENCY
Strike Team Leader (STEQ, STEN, STEQ)
Any higher position for which this position is a prerequisite

310-1 OTHER TRAINING WHICH SUPPORTS DEVELOPMENT OF KNOWLEDGE AND SKILLS
I-300 Intermediate Incident Command System
L-480 Fireline Leadership
S-336 Tactical Decision Making in Wildland Fire

27.1 - Exhibit 01--Continued

POSITION QUALIFICATIONS SECTIONS
Operations Positions

STRIKE TEAM LEADER CREW (STCR) - CONTINUED
Position Category: Incident Command System

FOREST SERVICE OTHER TRAINING WHICH SUPPORTS DEVELOPMENT OF KNOWLEDGE AND SKILLS
Geographic Area Intermediate Air Operations

Task Book available at:
http://www.nwcg.gov/pms/taskbook/operations/operation.htm

Note: The Task force Leader (TFLD) and Strike Team Leader (STCR, STEQ, STEN, STEQ) Position Task Books have been combined. However, the positions have not been combined. Strike Team Leader tasks are completed only once. The Required Experience for TFLD must be met prior to completing additional TFLD tasks.

Upon satisfactory performance in the prerequisite single resource boss position, the specific strike team leader task book may be initiated. Once qualified as a Strike Team Leader, any additional single resource boss qualifications will also qualify the individual in that corresponding strike team leader position – without having to complete the Strike Team Leader PTB for the new position – once agency Verification/certification is documented on the PTB certification page.

27.1 - Exhibit 01--Continued

POSITION QUALIFICATIONS SECTIONS
Operations Positions

CREW REPRESENTATIVE (CREP)
Position Category: Wildland Fire

310-1 REQUIRED TRAINING
RT-130 Annual Fireline Safety Refresher

FOREST SERVICE ADDITIONAL REQUIRED TRAINING
IS-700 National Incident Management System (NIMS), An Introduction

310-1 REQUIRED EXPERIENCE
Satisfactory performance as a Crew Boss, Single Resource (CRWB)
+
Successful position performance as a Crew Representative (CREP)

310-1 PHYSICAL FITNESS LEVEL
Moderate

310-1 OTHER POSITION ASSIGNMENTS THAT WILL MAINTAIN CURRENCY
Crew Boss, Single Resource (CRWB)
Interagency Resource Representative (IARR)

OTHER TRAINING WHICH SUPPORTS DEVELOPMENT OF KNOWLEDGE AND SKILLS
None

Task Book available at:
http://www.nwcg.gov/pms/taskbook/operations/operation.htm

27.1 - Exhibit 01--Continued

POSITION QUALIFICATIONS SECTIONS
Operations Positions

CREW BOSS (SINGLE RESOURCE) (CRWB)*
Position Category: Wildland Fire

310-1 REQUIRED TRAINING
S-290 Intermediate Wildland Fire Behavior
S-230 Crew Boss (Single Resource)
RT-130 Annual Fireline Safety Refresher

FOREST SERVICE ADDITIONAL REQUIRED TRAINING
I-200 Basic Incident Command System
IS-700 National Incident Management System (NIMS), An Introduction
L-280 Followership to Leadership
S-234 Ignition Operations
S-260 Interagency Incident Business Management
S-270 Basic Air Operations

310-1 REQUIRED EXPERIENCE
Satisfactory performance as a Firefighter Type 1 (FFT1)
        +
Successful position performance as a Crew Boss, Single Resource (CRWB) on a wildland fire incident

310-1 PHYSICAL FITNESS LEVEL
Arduous

310-1 OTHER POSITION ASSIGNMENTS THAT WILL MAINTAIN CURRENCY
Any single resource boss (HEQB, FELB, FIRB, ENGB, TRPB, HMGB)
Incident Commander Type 4 (ICT4)
Any higher position for which this position is a prerequisite

27.1 - Exhibit 01--Continued

POSITION QUALIFICATION SECTIONS
Operations Positions

CREW BOSS (SINGLE RESOURCE) (CRWB) - CONTINUED
Position Category: Wildland Fire

310-1 OTHER TRAINING WHICH SUPPORTS DEVELOPMENT OF KNOWLEDGE AND SKILLS
I-200 Basic Incident Command System
L-280 Followership to Leadership
S-270 Basic Air Operations
S-260 Interagency Incident Business Management
S-234 Ignition Operations

* The Position Task Book (PTB) for the Single Resource Boss positions differs from other PTBs. The first sets of tasks, common to all the Single Resource Boss positions, are the same as those required for the Crew Boss position. Additional specific tasks are required for the other types of Single Resource Boss positions (Engine, Dozer, Tractor/Plow, Felling, and Firing). When the PTB is issued to a trainee, the appropriate position(s) should be identified by crossing out the inappropriate positions on the cover. The trainee then needs to be signed off for all of the common tasks as well as those additional tasks that apply to the specific resource. Whether or not a qualified Single Resource Boss must re-complete the tasks common to all Single Resource Boss positions to become qualified as a Single Resource Boss for another resource is up to the discretion of the home unit (310-1,).

Task Book available at:
http://www.nwcg.gov/pms/taskbook/operations/operation.htm

27.1 - Exhibit 01--Continued

POSITION QUALIFICATIONS SECTIONS
Operations Positions

HEAVY EQUIPMENT BOSS (SINGLE RESOURCE) (HEQB)
Formerly known as Dozer Boss (DOZB)
Position Category: Wildland Fire

310-1 REQUIRED TRAINING
S-290 Intermediate Wildland Fire Behavior
S-230 Crew Boss (Single Resource)
RT-130 Annual Fireline Safety Refresher

FOREST SERVICE ADDITIONAL REQUIRED TRAINING
I-200 Basic Incident Command System
L-280 Followership to Leadership
S-232 Dozer Boss
S-234 Ignition Operations
S-260 Interagency Incident Business Management
S-270 Basic Air Operations
IS-700 National Incident Management System (NIMS), An Introduction

310-1 REQUIRED EXPERIENCE
Satisfactory performance as a Firefighter Type 1 (FFT1)
            +
Successful position performance as a Dozer Boss, Single Resource (HEQB) on a wildland fire incident

310-1 PHYSICAL FITNESS LEVEL
Arduous

310-1 OTHER POSITION ASSIGNMENTS THAT WILL MAINTAIN CURRENCY
Any single resource boss (CRWB, ENGB, FIRB, FELB, TRPB, HMGB)
Incident Commander Type 4 (ICT4)
Any higher position for which this position is a prerequisite

310-1 OTHER TRAINING WHICH SUPPORTS DEVELOPMENT OF KNOWLEDGE AND SKILLS
I-200 Basic Incident Command System
L-280 Followership to Leadership
S-270 Basic Air Operations
S-260 Interagency Incident Business Management
S-234 Ignition Operations
S-232 Dozer Boss (Single Resource)

Task Book available at:

http://www.nwcg.gov/pms/taskbook/operations/operation.htm

27.1 - Exhibit 01--Continued

POSITION QUALIFICATIONS SECTIONS
Operations Positions

FELLING BOSS (SINGLE RESOURCE) (FELB)
Position Category: Wildland Fire

310-1 REQUIRED TRAINING
S-290 Intermediate Wildland Fire Behavior
S-230 Crew Boss (Single Resource)
RT-130 Annual Fireline Safety Refresher

FOREST SERVICE ADDITIONAL REQUIRED TRAINING
I-200 Basic Incident Command System
IS-700 National Incident Management System (NIMS), An Introduction
L-280 Followership to Leadership
S-234 Ignition Operations
S-260 Interagency Incident Business Management
S-270 Basic Air Operations

FOREST SERVICE REQUIRED CERTIFICATION
Initial chain saw certification and triennial re-certification

FOREST SERVICE REQUIRED EXPERIENCE:
Currency as an unrestricted FALB or higher
        AND
Satisfactory performance as a Firefighter Type 1 (FFT1)
        AND
Successful position performance as a Felling Boss (FELB) on a wildfire or prescribed fire.

310-1 PHYSICAL FITNESS LEVEL
Arduous

310-1 OTHER POSITION ASSIGNMENTS THAT WILL MAINTAIN CURRENCY
Any single resource boss (CRWB, ENGB, FIRB, HEQB, TRPB, HMGB)
Incident Commander Type 4 (ICT4)
Any higher position for which this position is a prerequisite

27.1 - Exhibit 01--Continued

POSITION QUALIFICATIONS SECTIONS
Operations Positions

FELLING BOSS (SINGLE RESOURCE) (FELB)
Position Category: Wildland Fire

310-1 OTHER TRAINING WHICH SUPPORTS DEVELOPMENT OF KNOWLEDGE AND SKILLS

I-200 Basic Incident Command System

L-280 Followership to Leadership

S-270 Basic Air Operations

S-260 Interagency Incident Business Management

S-212 Wildland Fire Chain Saws

FOREST SERVICE OTHER TRAINING WHICH SUPPORTS DEVELOPMENT OF KNOWLEDGE AND SKILLS

Geographic Area Chainsaw Training

Task Book available at:

http://www.nwcg.gov/pms/taskbook/operations/operation.htm

27.1 - Exhibit 01--Continued

POSITION QUALIFICATIONS SECTIONS
Operations Positions

FIRING BOSS (SINGLE RESOURCE) (FIRB)
Position Category: Wildland Fire

310-1 REQUIRED TRAINING
S-290 Intermediate Wildland Fire Behavior
S-230 Crew Boss (Single Resource)
RT-130 Annual Fireline Safety Refresher

FOREST SERVICE ADDITIONAL REQUIRED TRAINING
I-200 Basic Incident Command System
IS-700 National Incident Management System (NIMS), An Introduction
L-280 Followership to Leadership
S-234 Ignition Operations
S-260 Interagency Incident Business Management
S-270 Basic Air Operations

FOREST SERVICE REQUIRED EXPERIENCE
Satisfactory performance as a Firefighter 1 (FFT1)
        AND
Successful position performance as a Firing Boss, Single Resource (FIRB) on a wildfire or prescribed fire.

310-1 PHYSICAL FITNESS LEVEL
Moderate

310-1 OTHER POSITION ASSIGNMENTS THAT WILL MAINTAIN CURRENCY
Any single resource boss (CRWB, ENGB, FELB, TRPB, HMGB, HEQB)
Incident Commander Type 4 (ICT4)
Any higher position for which this position is a prerequisite

Task Book available at:
    http://www.nwcg.gov/pms/taskbook/operations/operation.htm

27.1 - Exhibit 01--Continued

POSITION QUALIFICATIONS SECTIONS
Operations Positions

ENGINE BOSS (SINGLE RESOURCE) (ENGB)
Position Category: Wildland Fire

310-1 REQUIRED TRAINING
S-290 Intermediate Wildland Fire Behavior
S-230 Crew Boss (Single Resource)
RT-130 Annual Fireline Safety Refresher

FOREST SERVICE ADDITIONAL REQUIRED TRAINING:
I-200 Basic Incident Command System
IS-700 National Incident Management System (NIMS), An Introduction
L-280 Followership to Leadership
S-215 Fire Operations in the Urban Interface
S-231 Engine Boss or Geographic Area Engine Academy
S-234 Ignition Operations
S-260 Interagency Incident Business Management
S-270 Basic Air Operations

310-1 REQUIRED EXPERIENCE
Satisfactory performance as a Firefighter Type 1 (FFT1)
            +
Successful position performance as an Engine Boss, Single Resource (ENGB) on a wildland fire incident

310-1 PHYSICAL FITNESS LEVEL
Arduous

310-1 OTHER POSITION ASSIGNMENTS THAT WILL MAINTAIN CURRENCY
Any single resource boss (CRWB, HEQB, FIRB, FELB, TRPB, HMGB)
Incident Commander Type 4 (ICT4)
Any higher position for which this position is a prerequisite

27.1 - Exhibit 01--Continued

POSITION QUALIFICATIONS SECTIONS
Operations Positions

ENGINE BOSS (SINGLE RESOURCE) (ENGB) - CONTINUED
Position Category: Wildland Fire

310-1 OTHER TRAINING WHICH SUPPORTS DEVELOPMENT OF KNOWLEDGE AND SKILLS
I-200 Basic Incident Command System
L-280 Followership to Leadership
S-270 Basic Air Operations
S-260 Interagency Incident Business Management
S-234 Ignition Operations
S-231 Engine Boss (Single Resource)

Task Book available at:
http://www.nwcg.gov/pms/taskbook/operations/operation.htm

27.1 - Exhibit 01--Continued

POSITION QUALIFICATIONS SECTIONS
Operations Positions

HEAVY EQUIPMENT BOSS (SINGLE RESOURCE) (HEQB)
Formerly known as TRACTOR/PLOW BOSS (SINGLE RESOURCE) (TRPB)
Position Category: Wildland Fire

310-1 REQUIRED TRAINING
S-290 Intermediate Wildland Fire Behavior
S-230 Crew Boss (Single Resource)
RT-130 Annual Fireline Safety Refresher

FOREST SERVICE ADDITIONAL REQUIRED TRAINING:
I-200 Basic Incident Command System
IS-700 National Incident Management System (NIMS), An Introduction
L-280 Followership to Leadership
S-233 Tractor/Plow Boss
S-234 Ignition Operations
S-260 Interagency Incident Business Management
S-270 Basic Air Operations

310-1 REQUIRED EXPERIENCE
Satisfactory performance as a Firefighter Type 1 (FFT1)
        +
Successful position performance as a Tractor/Plow Boss, Single Resource (TRPB) on a wildland fire incident

310-1 PHYSICAL FITNESS LEVEL
Arduous

310-1 OTHER POSITION ASSIGNMENTS THAT WILL MAINTAIN CURRENCY
Any single resource boss (CRWB, ENGB, HEQB, FIRB, FELB, HMGB)
Incident Commander Type 4 (ICT4)
Any higher position for which this position is a prerequisite

310-1 OTHER TRAINING WHICH SUPPORTS DEVELOPMENT OF KNOWLEDGE AND SKILLS
I-200 Basic Incident Command System
L-280 Followership to Leadership
S-270 Basic Air Operations
S-260 Interagency Incident Business Management
S-233 Tractor/Plow Boss (Single Resource)

Task Book available at:
http://www.nwcg.gov/pms/taskbook/operations/operation.htm

27.1 - Exhibit 01--Continued

POSITION QUALIFICATIONS SECTIONS
Operations Positions

STAGING AREA MANAGER (STAM)
Position Category: Incident Command System

310-1 REQUIRED TRAINING
None

FOREST SERVICE ADDITIONAL REQUIRED TRAINING
I-100 Introduction to Incident Command System*
I-200 Basic Incident Command System
IS-700 National Incident Management System (NIMS), An Introduction
S-260 Interagency Incident Business Management

FOREST SERVICE REQUIRED EXPERIENCE
Desirable skills include record keeping, organizational abilities and communication skills
     AND
Successful position performance as a Staging Area Manager (STAM)

310-1 REQUIRED EXPERIENCE
Successful position performance as a Staging Area Manager (STAM)

310-1 PHYSICAL FITNESS LEVEL
Light

310-1 OTHER POSITION ASSIGNMENTS THAT WILL MAINTAIN CURRENCY
Firefighter Type 1 (FFT1)

310-1 OTHER TRAINING WHICH SUPPORTS DEVELOPMENT OF KNOWLEDGE AND SKILLS
I-200 Basic Incident Command System
J-236 Staging Area Manager

Task Book available at:
http://www.nwcg.gov/pms/taskbook/operations/operation.htm
*I-100 Online link: *http://training.nwcg.gov/classes/i100.htm*

27.1 - Exhibit 01--Continued

POSITION QUALIFICATIONS SECTIONS
Operations Positions

FIREFIGHTER TYPE 1 (FFT1)
Position Category: Wildland Fire

310-1 REQUIRED TRAINING
S-131 Firefighter Type 1
S-133 Look Up, Look Down, Look Around
RT-130 Annual Fireline Safety Refresher (annual requirement after certification in the position occurs)

FOREST SERVICE ADDITIONAL REQUIRED TRAINING
IS-700 National Incident Management System (NIMS), An Introduction
S-211 Portable Pumps and Water Use
S-212 Wildfire Power Saws

310-1 REQUIRED EXPERIENCE
Satisfactory performance as a Firefighter Type 2 (FFT2)
          +
Successful position performance as a Firefighter Type 1 (FFT1) on a wildland fire incident

310-1 PHYSICAL FITNESS LEVEL
Arduous

310-1 OTHER POSITION ASSIGNMENTS THAT WILL MAINTAIN CURRENCY
Incident Commander Type 5 (ICT5)
Any higher position for which this position is a prerequisite

310-1 OTHER TRAINING WHICH SUPPORTS DEVELOPMENT OF KNOWLEDGE AND SKILLS
S-212 Wildland Fire Chain Saws
S-211 Portable Pumps and Water Use

Task Book available at:
http://www.nwcg.gov/pms/taskbook/operations/operation.htm

Note: The Firefighter Type 1 (FFT1) and Incident Commander Type 5 (ICT5) Position Task Books have been combined. However, the positions have not been combined. The FFT1 tasks are completed only once; additional tasks must be completed to meet the ICT5 level. The FFT1 and ICT5 tasks can be completed simultaneously. The Required Experience is satisfactory performance as a Firefighter Type 2 (FFT2).

<u>**27.1 - Exhibit 01--Continued**</u>

POSITION QUALIFICATIONS SECTIONS
Operations Positions

FIREFIGHTER TYPE 2 (FFT2)
Position Category: Wildland Fire

310-1 REQUIRED TRAINING
Basic Firefighter Training:
       I-100 Introduction to Incident Command System*
       L-180 Human Factors on the Fireline
       S-190 Introduction to Wildland Fire Behavior
       S-130 Firefighting Training
RT-130 Annual Fireline Safety Refresher**
FOREST SERVICE ADDITIONAL REQUIRED TRAINING
IS-700 National Incident Management System (NIMS), An Introduction

310-1 REQUIRED EXPERIENCE
None

310-1 PHYSICAL FITNESS LEVEL
Arduous

310-1 OTHER POSITION ASSIGNMENTS THAT WILL MAINTAIN CURRENCY
Any higher position for which this position is a prerequisite

310-1 OTHER TRAINING WHICH SUPPORTS DEVELOPMENT OF KNOWLEDGE AND SKILLS
None

Note: For the FFT2 position, satisfactory completion of the required training meets the position qualification requirements. This position does not require completion of a Position Task Book.
*I-100 Online link: *http://training.nwcg.gov/classes/i100.htm*
**Note: Annual Fireline Safety Refresher (RT-130) is not required for the first year as a Firefighter Type 2 (FFT2); however, it is required for subsequent years.

27.1 - Exhibit 01--Continued

POSITION QUALIFICATIONS SECTIONS
Air Operations Positions

AREA COMMAND AVIATION COORDINATOR (ACAC)
Position Category: Incident Command System

310-1 REQUIRED TRAINING
S-620 Area Command

FOREST SERVICE ADDITIONAL REQUIRED TRAINING
IS-700 National Incident Management System (NIMS), An Introduction
IS-800 National Response Plan (NRP), An Introduction

310-1 REQUIRED EXPERIENCE
Satisfactory performance as an Air Operations Branch Director (AOBD) on a National Type 1
Incident Management Team
            +
Successful position performance as an Area Command Aviation Coordinator (ACAC) on a
wildfire incident

310-1 PHYSICAL FITNESS LEVEL
None Required

310-1 OTHER POSITION ASSIGNMENTS THAT WILL MAINTAIN CURRENCY
Air Operations Branch Director (AOBD)

**310-1 OTHER TRAINING WHICH SUPPORTS DEVELOPMENT OF KNOWLEDGE
AND SKILLS**
None

Task Book available at:
http://www.nwcg.gov/pms/taskbook/air/air.htm

27.1 - Exhibit 01--Continued

POSITION QUALIFICATIONS SECTIONS
Air Operations Positions

AIR OPERATIONS BRANCH DIRECTOR (AOBD)
Position Category: Incident Command System

310-1 REQUIRED TRAINING
Air Operations Branch Director (S-470)

FOREST SERVICE ADDITIONAL REQUIRED TRAINING
IS-700 National Incident Management System (NIMS), An Introduction
IS-800 National Response Plan (NRP), An Introduction

310-1 REQUIRED EXPERIENCE
Satisfactory performance as an Air Support Group Supervisor (ASGS)
　　　+
Successful position performance as an Air Operations Branch Director (AOBD) on a wildfire incident

310-1 PHYSICAL FITNESS LEVEL
None Required

310-1 OTHER POSITION ASSIGNMENTS THAT WILL MAINTAIN CURRENCY
Air Support Group Supervisor (ASGS)
Any higher position for which this position is a prerequisite

310-1 OTHER TRAINING WHICH SUPPORTS DEVELOPMENT OF KNOWLEDGE AND SKILLS
I-400 Advanced Incident Command System
L-480 Organizational Leadership in the Fire Service

27.1 - Exhibit 01--Continued

POSITION QUALIFICATIONS SECTIONS
Air Operations Positions

AIR OPERATIONS BRANCH DIRECTOR (AOBD) - CONTINUED
Position Category: Incident Command System

FOREST SERVICE OTHER TRAINING WHICH SUPPORTS DEVELOPMENT OF KNOWLEDGE AND SKILLS
S-378 Air Tactical Group Supervisor
A-101 Aviation Safety
A-103 FAA NOTAM Administration
A-105 Aviation Life Support Equipment
A-106 Aviation Mishap Reporting
A-107 Aviation Policy and Regulations I
A-109 Aviation Radio Use
A-112 Mission Planning and Flight Request Process
A-113 Crash Survival
A-200 Mishap Review
A-202 Interagency Aviation Organizations
A-203 Basic Airspace
A-204 Aircraft Capabilities and Limitations
A-301 Implementing Aviation Safety and Accident Prevention
A-302 Personal Responsibility and Liability
A-303 Human Factors in Aviation
A-305 Risk Management
A-307 Aviation Policy and Regulations II

Task Book available at:
http://www.nwcg.gov/pms/taskbook/air/air.htm

27.1 - Exhibit 01--Continued

POSITION QUALIFICATIONS SECTIONS
Air Operations Positions

AIR SUPPORT GROUP SUPERVISOR (ASGS)
Position Category: Incident Command System

310-1 REQUIRED TRAINING
RT-130 Annual Fireline Safety Refresher

FOREST SERVICE ADDITIONAL REQUIRED TRAINING
S-375 Air Support Group Supervisor
IS-700 National Incident Management System (NIMS), An Introduction

310-1 REQUIRED EXPERIENCE
Satisfactory performance as a Helibase Manager Type 1(HEB1)
         +
Successful position performance as an Air Support Group Supervisor (ASGS) on a wildland fire incident

310-1 PHYSICAL FITNESS LEVEL
None Required

310-1 OTHER POSITION ASSIGNMENTS THAT WILL MAINTAIN CURRENCY
Helibase Manager Type 1 (HEB1)
Any higher position for which this position is a prerequisite

310-1 OTHER TRAINING WHICH SUPPORTS DEVELOPMENT OF KNOWLEDGE AND SKILLS
I-300 Intermediate Incident Command System
S-375 Air Support Group Supervisor

27.1 - Exhibit 01--Continued

POSITION QUALIFICATIONS SECTIONS
Air Operations Positions

AIR SUPPORT GROUP SUPERVISOR (ASGS) - CONTINUED
Position Category: Incident Command System

FOREST SERVICE OTHER TRAINING WHICH SUPPORTS DEVELOPMENT OF KNOWLEDGE AND SKILLS
Geographic Area Intermediate Air Operations
A-101 Aviation Safety
A-103 FAA NOTAM System
A-105 Aviation Life Support Equipment
A-106 Aviation Mishap Reporting
A-107 Aviation Policy and Regulations 1
A-109 Aviation Radio Use
A-112 Mission Planning and Flight Request Process
A-113 Crash Survival
A-200 Mishap Review
A-202 Interagency Aviation Organizations
A-203 Basic Airspace
A-204 Aircraft Capabilities and Limitations

Task Book available at:
http://www.nwcg.gov/pms/taskbook/air/air.htm

27.1 - Exhibit 01--Continued

POSITION QUALIFICATIONS SECTIONS
Air Operations Positions

HELIBASE MANAGER TYPE 1 (6 OR MORE HELICOPTERS) (HEB1)
Position Category: Incident Command System

310-1 REQUIRED TRAINING
RT-130 Annual Fireline Safety Refresher

FOREST SERVICE ADDITIONAL REQUIRED TRAINING
A-110 Aviation Transportation of Hazardous Materials (Must attend every three years)
IS-700 National Incident Management System (NIMS), An Introduction

310-1 REQUIRED EXPERIENCE
Satisfactory performance as a Helibase Manager Type 2 (HEB2)
        +
Successful position performance as a Helibase Manager Type 1 (HEB1)

310-1 PHYSICAL FITNESS LEVEL
Light

310-1 OTHER POSITION ASSIGNMENTS THAT WILL MAINTAIN CURRENCY
Helibase Manager Type 2 (HEB2)
Any higher position for which this position is a prerequisite

OTHER TRAINING WHICH SUPPORTS DEVELOPMENT OF KNOWLEDGE AND SKILLS
None

Task Book available at:
http://www.nwcg.gov/pms/taskbook/air/air.htm

27.1 - Exhibit 01--Continued

POSITION QUALIFICATIONS SECTIONS
Air Operations Positions

HELIBASE MANAGER 2 (ONE TO FIVE HELICOPTERS) (HEB2)
Position Category: Incident Command System

310-1 REQUIRED TRAINING
Helibase Manager (S-371)
Annual Fireline Safety Refresher (RT-130)

FOREST SERVICE ADDITIONAL REQUIRED TRAINING
A-110 Aviation Transportation of Hazardous Materials (Must attend every three years)
IS-700 National Incident Management System (NIMS), An Introduction

310-1 REQUIRED EXPERIENCE
Satisfactory performance as a Helicopter Manager (HMGB)

+ .

Successful position performance as a Helibase Manager Type 2 (HEB2)

310-1 PHYSICAL FITNESS LEVEL
Light

310-1 OTHER POSITION ASSIGNMENTS THAT WILL MAINTAIN CURRENCY
Helicopter Manager (HMGB)
Any higher position for which this position is a prerequisite

310-1 OTHER TRAINING WHICH SUPPORTS DEVELOPMENT OF KNOWLEDGE AND SKILLS
I-300 Intermediate Incident Command System
L-380 Fireline Leadership

Task Book available at:
http://www.nwcg.gov/pms/taskbook/air/air.htm

POSITION QUALIFICATIONS SECTIONS
Air Operations Positions

HELICOPTER MANAGER - (HMGB)
Position Category: Incident Support

310-1 REQUIRED TRAINING
S-230 Crew Boss (Single Resource)
S-290 Intermediate Wildland Fire Behavior
RT-372 Helicopter Manager Workshop (Triennial refresher, after completion of S-372 must attend RT-372 within three years and every three years thereafter)
S-372 Helicopter Management
RT-130 Annual Fireline Safety Refresher

FOREST SERVICE ADDITIONAL REQUIRED TRAINING:
A-110 Aviation Transportation of Hazardous Materials (Must attend every three years)
A-219 Helicopter Transport of External Cargo
RT-219FS Helicopter Transport of External Cargo (Must attend A-219 within three years and every three years thereafter)
IS-700 National Incident Management System (NIMS), An Introduction
I-200 Basic Incident Command System
L-280 Followership to Leadership
S-234 Ignition Operations
S-260 Interagency Incident Business Management
S-270 Basic Air Operations
Aviation Business System Training**
Contract Administration Training (Agency Specific)

310-1 REQUIRED EXPERIENCE
Satisfactory performance as a Helicopter Crewmember (HECM)

    +

Satisfactory performance as a Firefighter Type 1 (FFT1)

    +

Successful position performance as a Helicopter Manager (HMGB) on a wildland fire incident

27.1 - Exhibit 01--Continued

POSITION QUALIFICATIONS SECTIONS
Air Operations Positions

HELICOPTER MANAGER - (HMGB) – CONTINUED
Position Category: Incident Support

310-1 PHYSICAL FITNESS LEVEL
Moderate

310-1 OTHER POSITION ASSIGNMENTS THAT WILL MAINTAIN CURRENCY
Any Single Resource Boss (CRWB, HEQB, ENGB, FELB, FIRB, TRPB)
Any higher position for which this position is a prerequisite

FOREST SERVICE OTHER POSITION ASSIGNMENTS THAT WILL MAINTAIN CURRENCY
HMGB
Any higher position for which this position is a prerequsite

310-1 OTHER TRAINING WHICH SUPPORTS DEVELOPMENT OF KNOWLEDGE AND SKILLS
None

Task Book available at:
http://www.nwcg.gov/pms/taskbook/air/air.htm

*Note: When Helicopter Managers are intended to be utilized for other missions they must be ordered with additional qualifications such as: ICT4, PLDO, Agency Exclusive Use Prerequisites, etc.
**Online training at: *http://www.fs.fed.us/business/abs/training.php*

27.1 - Exhibit 01--Continued

POSITION QUALIFICATIONS SECTIONS
Air Operations Positions

HELICOPTER CREWMEMBER (HECM)
Position Category: Wildland Fire

310-1 REQUIRED TRAINING
S-271 Helicopter Crewmember
RT-130 Annual Fireline Safety Refresher

FOREST SERVICE ADDITIONAL REQUIRED TRAINING
A-110 Aviation Transportation of Hazardous Materials (Must attend every three years)
A-219 Helicopter Transport of External Cargo
RT-219FS Helicopter Transport of External Cargo (Must attend A-219 within three years and every three years thereafter)
IS-700 National Incident Management System (NIMS), An Introduction
L-180 Human Factors on the Fireline (if not obtained in S-130 2003 revision)

310-1 REQUIRED EXPERIENCE
Satisfactory performance as a Firefighter Type 2 (FFT2)
           +
Successful position performance as a Helicopter Crewmember (HECM)

310-1 PHYSICAL FITNESS LEVEL
Arduous

310-1 OTHER POSITION ASSIGNMENTS THAT WILL MAINTAIN CURRENCY
Any higher position for which this position is a prerequisite

310-1 OTHER TRANIING WHICH SUPPORTS DEVELOPMENT OF KNOWLEDGE AND SKILLS
None

Task Book available at:
http://www.nwcg.gov/pms/taskbook/air/air.htm

27.1 - Exhibit 01--Continued

POSITION QUALIFICATIONS SECTIONS
Air Operations Positions

AIR TACTICAL GROUP SUPERVISOR (ATGS)
Position Category: Incident Command System

310-1 REQUIRED TRAINING:
S-378 Air Tactical Group Supervisor
RT-130 Annual Fireline Safety Refresher

FOREST SERVICE ADDITIONAL REQUIRED TRAINING:
IS-700 National Incident Management System (NIMS), An Introduction
IS-800 National Response Plan (NRP), An Introduction

310-1 REQUIRED EXPERIENCE:
Satisfactory performance as a Division Group Supervisor (DIVS)

+

Successful position performance as an Air Tactical Group Supervisor (ATGS) on a wildfire incident

OR

Satisfactory performance as an Incident Commander Type 3 (ICT3)

+

Successful position performance as an Air Tactical Group Supervisor (ATGS) on a wildfire incident

310-1 PHYSICAL FITNESS LEVEL
None required

310-1 OTHER POSITION ASSIGNMENTS THAT WILL MAINTAIN CURRENCY
None

310-1 OTHER TRAINING WHICH SUPPORTS DEVELOPMENT OF KNOWLEDGE AND SKILLS
None

27.1 - Exhibit 01--Continued

POSITION QUALIFICATIONS SECTIONS
Air Operations Positions

AIR TACTICAL GROUP SUPERVISOR (ATGS) - CONTINUED
Position Category: Incident Command System

FOREST SERVICE OTHER TRAINING WHICH SUPPORTS DEVELOPMENT OF KNOWLEDGE AND SKILLS:
Aerial Retardant Application and Use
A-101 Aviation Safety (All Aircraft)
A-103 FAA NOTAM System
A-105 Aviation Life Support Equipment
A-106 Aviation Mishap Reporting
A-107 Aviation Policy and Regulations I
A-109 Aviation Radio Use
A-112 Mission Planning and Flight Request Process
A-113 Crash Survival
A-200 Mishap Review--
A-202 Interagency Aviation Organizations
A-203 Basic Airspace
A-204 Aircraft Capabilities and Limitations
A-206 Aviation Acquisition and Procurement
A-301 Implementing Aviation Safety and Accident Prevention
A-302 Personal Responsibility and Liability
A-303 Human Factors in Aviation
A-305 Risk Management
A-307 Aviation Policy and Regulations II
A-311 Aviation Planning

Reference materials for this position are contained in the Interagency Aerial Supervision Guide

Task Book available at:
http://www.nwcg.gov/pms/taskbook/air/air.htm

<u>**27.1 - Exhibit 01--Continued**</u>

POSITION QUALIFICATIONS SECTIONS
Air Operations Positions

AIR TANKER/FIXED WING COORDINATOR (ATCO)
Position Category: Incident Command System

310-1 REQUIRED TRAINING:
Agency established

FOREST SERVICE ADDITIONAL REQUIRED TRAINING
I-100 Introduction to Incident Command System*
IS-700 National Incident Management System (NIMS), An Introduction
S-190 Introduction to Wildland Fire Behavior
S-270 Basic Air Operations
S-290 Intermediate Fire Behavior
S-336 Fire Suppression Tactics or equivalent training (see chapter 30 of this Handbook)
S-378 Air Tactical Group Supervisor

310-1 REQUIRED EXPERIENCE
Agency established

310-1 PHYSICAL FITNESS LEVEL
None required

310-1 OTHER POSITION ASSIGNMENTS THAT WILL MAINTAIN CURRENCY
None

310-1 OTHER TRAINING WHICH SUPPORTS DEVELOPMENT OF KNOWLEDGE AND SKILLS
L-380 Fireline Leadership

*I-100 Online link: *http://training.nwcg.gov/classes/i100.htm*

27.1 - Exhibit 01--Continued

POSITION QUALIFICATIONS SECTIONS
Air Operations Positions

AIR TANKER/FIXED WING COORDINATOR (ATCO) - CONTINUED
Position Category: Incident Command System

FOREST SERVICE OTHER TRAINING WHICH SUPPORTS DEVELOPMENT OF KNOWLEDGE AND SKILLS
Geographic Area Intermediate Air Operations
I-200 Basic Incident Command System
A-101 Aviation Safety (All Aircraft)
A-103 FAA NOTAM System
A-105 Aviation Life Support Equipment
A-106 Aviation Mishap Reporting
A-107 Aviation Policy and Regulations I

A-112 Mission Planning and Flight Request Process
A-113 Crash Survival
A-200 Mishap Review
A-202 Interagency Aviation Organizations
A-203 Basic Airspace
A-204 Aircraft Capabilities and Limitations
A-206 Aviation Acquisition and Procurement
A-301 Implementing Aviation Safety and Accident Prevention
A-302 Personal Responsibility and Liability
A-303 Human Factors in Aviation
A-305 Risk Management
A-307 Aviation Policy and Regulations II
A-311 Aviation Planning

POSITION QUALIFICATIONS SECTIONS
Air Operations Positions

HELICOPTER COORDINATOR (HLCO)
Position Category: Incident Command System

310-1 REQUIRED TRAINING
S-378 Air Tactical Group Supervisor
RT-130 Annual Fireline Safety Refresher

FOREST SERVICE ADDITIONAL REQUIRED TRAINING
IS-700 National Incident Management System (NIMS), An Introduction
IS-800 National Response Plan (NRP), An Introduction

310-1 REQUIRED EXPERIENCE
Satisfactory performance as a Task Force Leader (TFLD)
    +
Successful position performance as a Helicopter Coordinator (HLCO)
    OR
Satisfactory performance in one Strike Team Leader position (STCR, STEN, STEQ, STEQ)
    +
Successful position performance as Helicopter Coordinator (HLCO)

310-1 PHYSICAL FITNESS LEVEL
None required

310-1 OTHER POSITION ASSIGNMENTS THAT WILL MAINTAIN CURRENCY
Air Tactical Group Supervisor (ATGS)

310-1 OTHER TRAINING WHICH SUPPORTS DEVELOPMENT OF KNOWLEDGE AND SKILLS
None

Task Book available at:
http://www.nwcg.gov/pms/taskbook/air/air.htm

27.1 - Exhibit 01--Continued

POSITION QUALIFICATIONS SECTIONS
Air Operations Positions

SINGLE ENGINE AIR TANKER MANAGER (SEMG)
Position Category: Incident Support

310-1 REQUIRED TRAINING
S-273 Single Engine Air Tanker Manager
S-270 Basic Air Operations

FOREST SERVICE ADDITIONAL REQUIRED TRAINING
I-100 Introduction to Incident Command System*
IS-700 National Incident Management System (NIMS), An Introduction

FOREST SERVICE CERTIFICATION
RT-273 Single Engine Air Tanker Manager Workshop (triennial)

310-1 PHYSICAL FITNESS LEVEL
None required

310-1 OTHER POSITION ASSIGNMENTS THAT WILL MAINTAIN CURRENCY
Helicopter Manager (HMGB)
Air Tanker Base Manager (ATBM)
Fixed-Wing Base Manager (FWBM)

310-1 OTHER TRAINING WHICH SUPPORTS DEVELOPMENT OF KNOWLEDGE AND SKILLS
D-110 Dispatch Recorder
I-200 Basic Incident Command System
Basic Firefighter Training:
    L-180 Human Factors on the Fireline
    S-130 Firefighter Training
    S-190 Introduction to Wildland Fire Behavior

Task Book available at:
http://www.nwcg.gov/pms/taskbook/air/air.htm
*I-100 Online link: *http://training.nwcg.gov/classes/i100.htm*

27.1 - Exhibit 01--Continued

POSITION QUALIFICATIONS SECTIONS
Air Operations Positions

DECK COORDINATOR (DECK)
Position Category: Incident Support

310-1 REQUIRED TRAINING
None

FOREST SERVICE ADDITIONAL REQUIRED TRAINING
A-110 Aviation Transportation of Hazardous Materials (Must attend every three years)
A-219 Helicopter Transport of External Cargo
RT-219FS Helicopter Transport of External Cargo (Must attend A-219 within three years and every three years thereafter)
IS-700 National Incident Management System (NIMS), An Introduction

310-1 REQUIRED EXPERIENCE
Satisfactory performance as a Helicopter Crewmember (HECM)
+
Satisfactory performance as a Takeoff and Landing Coordinator (TOLC)
+
Successful position performance as a Deck Coordinator (DECK)

310-1 PHYSICAL FITNESS LEVEL
Light

310-1 OTHER POSITION ASSIGNMENTS THAT WILL MAINTAIN CURRENCY
Helibase Manager Type 2 (HEB2)
Takeoff and Landing Coordinator (TOLC)

310-1 OTHER TRAINING WHICH SUPPORTS DEVELOPMENT OF KNOWLEDGE AND SKILLS
None
Task Book available at:
http://www.nwcg.gov/pms/taskbook/air/air.htm

27.1 - Exhibit 01--Continued

POSITION QUALIFICATIONS SECTIONS
Air Operations Positions

TAKEOFF AND LANDING COORDINATOR (TOLC)
Position Category: Incident Support

310-1 REQUIRED TRAINING
None

FOREST SERVICE ADDITIONAL REQUIRED TRAINING
IS-700 National Incident Management System (NIMS), An Introduction

310-1 REQUIRED EXPERIENCE
Satisfactory performance as Aircraft Base Radio Operator (ABRO)
        +
Successful performance as a Takeoff and Landing Coordinator (TOLC)

310-1 PHYSICAL FITNESS LEVEL
Light

310-1 OTHER POSITION ASSIGNMENTS THAT WILL MAINTAIN CURRENCY
Helibase Manager Type 2 (HEB2)
Aircraft Base Radio Operator (ABRO)
Any higher position for which this position is a prerequisite

310-1 OTHER TRAINING WHICH SUPPORTS DEVELOPMENT OF KNOWLEDGE AND SKILLS
None

Task Book available at:
http://www.nwcg.gov/pms/taskbook/air/air.htm

27.1 - Exhibit 01--Continued

POSITION QUALIFICATIONS SECTIONS
Air Operations Positions

AIRCRAFT BASE RADIO OPERATOR (ABRO)
Position Category: Incident Support

310-1 REQUIRED TRAINING
None

FOREST SERVICE ADDITIONAL REQUIRED TRAINING
I-100 Introduction to Incident Command System*
IS-700 National Incident Management System (NIMS), An Introduction

310-1 REQUIRED EXPERIENCE
Successful position performance as an Aircraft Base Radio Operator (ABRO)

310-1 PHYSICAL FITNESS LEVEL
None required

310-1 OTHER POSITION ASSIGNMENTS THAT WILL MAINTAIN CURRENCY
Radio Operator (RADO)
Helicopter Crewmember (HECM)
Any higher position for which this position is a prerequisite

310-1 OTHER TRAINING WHICH SUPPORTS DEVELOPMENT OF KNOWLEDGE AND SKILLS
None

FOREST SERVICE OTHER TRAINING WHICH SUPPORTS DEVELOPMENT OF KNOWLEDGE AND SKILLS
A-101 Aviation Safety (All Aircraft)
A-109 Aviation Radio Use

Task Book available at:
http://www.nwcg.gov/pms/taskbook/air/air.htm
*I-100 Online link: http://training.nwcg.gov/classes/i100.htm

<u>**27.1 - Exhibit 01--Continued**</u>

POSITION QUALIFICATIONS SECTIONS
Planning Positions

ASSISTANT AREA COMMANDER, PLANNING (ACPC)
Position Category: Incident Command System

310-1 REQUIRED TRAINING
S-620 Area Command

FOREST SERVICE ADDITIONAL REQUIRED TRAINING
IS-700 National Incident Management System (NIMS), An Introduction
IS-800 National Response Plan (NRP), An Introduction

310-1 REQUIRED EXPERIENCE
Satisfactory performance as an Incident Commander or General Staff on a national Type 1
Incident Management Team
        +
Successful position performance as an Assistant Area Commander, Planning (ACPC) on a
wildland fire incident

310-1 PHYSICAL FITNESS LEVEL
None

310-1 OTHER POSITION ASSIGNMENTS THAT WILL MAINTAIN CURRENCY
Assistant Area Commander, Logistics (ACLC)
Incident Commander Type 1 (ICT1)
Any higher position for which this position is a prerequisite

**310-1 OTHER TRAINING WHICH SUPPORTS DEVELOPMENT OF KNOWLEDGE
AND SKILLS**
None

Task Book available at:
http://www.nwcg.gov/pms/taskbook/planning/planning.htm

27.1 - Exhibit 01--Continued

POSITION QUALIFICATIONS SECTIONS
Planning Positions

PLANNING SECTION CHIEF TYPE 1 (PSC1)
Position Category: Incident Command System

310-1 REQUIRED TRAINING
S-520 Advanced Incident Management

FOREST SERVICE ADDITIONAL REQUIRED TRAINING
IS-700 National Incident Management System (NIMS), An Introduction
IS-800 National Response Plan (NRP), An Introduction

310-1 REQUIRED EXPERIENCE
Satisfactory performance as a Planning Section Chief Type 2 (PSC2)

+

Successful position performance as a Planning Section Chief Type 1 (PSC1) on a wildland fire incident

310-1 PHYSICAL FITNESS LEVEL
None required

310-1 OTHER POSITION ASSIGNMENTS THAT WILL MAINTAIN CURRENCY
Planning Section Chief Type 2 (PSC2)
Any higher position for which this position is a prerequisite

310-1 OTHER TRAINING WHICH SUPPORTS DEVELOPMENT OF KNOWLEDGE AND SKILLS
None

Task Book available at:
http://www.nwcg.gov/pms/taskbook/planning/planning.htm

27.1 - Exhibit 01--Continued

POSITION QUALIFICATIONS SECTIONS
Planning Positions

PLANNING SECTION CHIEF TYPE 2 (PSC2)
Position Category: Incident Command System

310-1 REQUIRED TRAINING
S-420 Command and General Staff

FOREST SERVICE ADDITIONAL REQUIRED TRAINING
IS-700 National Incident Management System (NIMS), An Introduction
IS-800 National Response Plan (NRP), An Introduction

310-1 REQUIRED EXPERIENCE
Satisfactory performance as a Situation Unit Leader (SITL)

+

Satisfactory performance as a Resource Unit Leader (RESL)

+

Successful position performance as a Planning Section Chief Type 2 (PSC2) on a wildland fire incident

310-1 PHYSICAL FITNESS LEVEL
None required

310-1 OTHER POSITION ASSIGNMENTS THAT WILL MAINTAIN CURRENCY
Situation Unit Leader (SITL)
Resource Unit Leader (RESL)
Any higher position for which this position is a prerequisite

310-1 OTHER TRAINING WHICH SUPPORTS DEVELOPMENT OF KNOWLEDGE AND SKILLS
I-400 Advanced Incident Command System
L-480 Organizational Leadership in the Fire Service
S-440 Planning Section Chief

Task Book available at:
http://www.nwcg.gov/pms/taskbook/planning/planning.htm

27.1 - Exhibit 01--Continued

POSITION QUALIFICATIONS SECTIONS
Planning Positions

SITUATION UNIT LEADER (SITL)
Position Category: Incident Command System

310-1 REQUIRED TRAINING
RT-130 Annual Fireline Safety Refresher

FOREST SERVICE ADDITIONAL REQUIRED TRAINING
IS-700 National Incident Management System (NIMS), An Introduction
IS-800 National Response Plan (NRP), An Introduction

310-1 REQUIRED EXPERIENCE
Satisfactory performance as an Incident Commander Type 4 (ICT4)
    +
Successful position performance as a Situation Unit Leader (SITL) on a wildland fire incident

310-1 PHYSICAL FITNESS LEVEL
Light

310-1 OTHER POSITION ASSIGNMENTS THAT WILL MAINTAIN CURRENCY
Field Observer (FOBS)
Any higher position for which this position is a prerequisite

310-1 OTHER TRAINING WHICH SUPPORTS DEVELOPMENT OF KNOWLEDGE AND SKILLS
I-300 Intermediate Incident Command System
S-346 Situation Unit Leader
L-380 Fireline Leadership

Task Book available at:
http://www.nwcg.gov/pms/taskbook/planning/planning.htm

27.1 - Exhibit 01--Continued

POSITION QUALIFICATIONS SECTIONS
Planning Positions

RESOURCE UNIT LEADER (RESL)
Position Category: Incident Command System

310-1 REQUIRED TRAINING
None

FOREST SERVICE ADDITIONAL REQUIRED TRAINING
IS-700 National Incident Management System (NIMS), An Introduction
IS-800 National Response Plan (NRP), An Introduction

310-1 REQUIRED EXPERIENCE
Satisfactory performance as a Status/Check-in Recorder (SCKN)
   +
Successful position performance as a Resource Unit Leader

310-1 PHYSICAL FITNESS LEVEL
None required

310-1 OTHER POSITION ASSIGNMENTS THAT WILL MAINTAIN CURRENCY
Demobilization Unit Leader (DMOB)
Status/Check-In Recorder (SCKN)
Any higher position for which this position is a prerequisite

310-1 OTHER TRAINING WHICH SUPPORTS DEVELOPMENT OF KNOWLEDGE AND SKILLS
I-300 Intermediate Incident Command System
L-380 Fireline Leadership
S-260 Interagency Incident Business Management
S-349 Resource Unit Leader and Demobilization Unit Leader

Task Book available at:
http://www.nwcg.gov/pms/taskbook/planning/planning.htm

<u>**27.1 - Exhibit 01--Continued**</u>

POSITION QUALIFICATIONS SECTIONS
Planning Positions

STATUS/CHECK-IN RECORDER (SCKN)
Position Category: Incident Command System

310-1 REQUIRED TRAINING
None

FOREST SERVICE ADDITIONAL REQUIRED TRAINING
I-100 Introduction to Incident Command System*
IS-700 National Incident Management System (NIMS), An Introduction
S-248 Status/Check-In Recorder

FOREST SERVICE CERTIFICATION
On-the-job experience with Incident Base Automation (I-SUITE)

310-1 REQUIRED EXPERIENCE
Successful position performance as a Status/Check-in Recorder (SCKN)

310-1 PHYSICAL FITNESS LEVEL
None required

310-1 OTHER POSITION ASSIGNMENTS THAT WILL MAINTAIN CURRENCY
Any higher position for which this position is a prerequisite

310-1 OTHER TRAINING WHICH SUPPORTS DEVELOPMENT OF KNOWLEDGE AND SKILLS
I-Suite Incident Base Automation
I-100 Introduction to Incident Command System
L-180 Human Factors on the Fireline
S-248 Status/Check-In Recorder
S-110 Basic Wildland Fire Orientation

Task Book available at:
http://www.nwcg.gov/pms/taskbook/planning/planning.htm
*I-100 Online link: *http://training.nwcg.gov/classes/i100.htm*

27.1 - Exhibit 01--Continued

POSITION QUALIFICATIONS SECTIONS
Planning Positions

DOCUMENTATION UNIT LEADER (DOCL)
Position Category: Incident Command System

310-1 REQUIRED TRAINING
None

FOREST SERVICE ADDITIONAL REQUIRED TRAINING
I-100 Introduction to Incident Command System*
IS-700 National Incident Management System (NIMS), An Introduction
IS-800 National Response Plan (NRP), An Introduction

310-1 PHYSICAL FITNESS LEVEL
None Required

310-1 OTHER POSITION ASSIGNMENTS THAT WILL MAINTAIN CURRENCY
Planning Section Chief Type 2 (PSC2)

310-1 OTHER TRAINING WHICH SUPPORTS DEVELOPMENT OF KNOWLEDGE AND SKILLS
I-300 Intermediate Incident Command System
J-342 Documentation Unit Leader
S-110 Basic Wildland Fire Orientation

Task Book available at:
http://www.nwcg.gov/pms/taskbook/planning/planning.htm
*I-100 Online link: *http://training.nwcg.gov/classes/i100.htm*

27.1 - Exhibit 01--Continued

POSITION QUALIFICATIONS SECTIONS
Planning Positions

DEMOBILIZATION UNIT LEADER (DMOB)
Position Category: Incident Command System

310-1 REQUIRED TRAINING
None

FOREST SERVICE ADDITIONAL REQUIRED TRAINING
IS-700 National Incident Management System (NIMS), An Introduction
IS-800 National Response Plan (NRP), An Introduction

310-1 REQUIRED EXPERIENCE
Satisfactory performance as a Resource Unit Leader (RESL)

    +

Successful position performance as a Demobilization Unit Leader (DMOB)

310-1 PHYSCIAL FITNESS LEVEL
None Required

310-1 OTHER POSITION ASSIGNMENTS THAT WILL MAINTAIN CURRENCY
Planning Section Chief Type 2 (PSC2)
Resource Unit Leader (RESL)
Expanded Dispatch Support Dispatcher (EDSD)

310-1 OTHER TRAINING WHICH SUPPORTS DEVELOPMENT OF KNOWLEDGE AND SKILLS
S-349 Resources Unit Leader and Demobilization Unit Leader

Task Book available at:
http://www.nwcg.gov/pms/taskbook/planning/planning.htm

27.1 - Exhibit 01--Continued

POSITION QUALIFICATIONS SECTIONS
Planning Positions

STRATEGIC OPERATIONAL PLANNER (SOPL)
Position Category: Wildland Fire

310-1 REQUIRED TRAINING
S-580 Advanced Fire Use Applications
RT-130 Annual Fireline Safety Refresher   OR
S482 Advanced Fire Management Applications
RT-130 Annual Fireline Safety Refresher

FOREST SERVICE ADDITIONAL REQUIRED TRAINING:
IS-700 National Incident Management System (NIMS), An Introduction
IS-800 National Response Plan (NRP), An Introduction

310-1 REQUIRED EXPERIENCE
Satisfactory performance as a Prescribed Fire Burn Boss Type 2 (RXB2)
+
Satisfactory performance as an Incident Commander Type 3 (ICT3)
+
Successful position performance as a Strategic Operational Planner (SOPL) on a wildfire

OR

Satisfactory performance as a Prescribed Fire Burn Boss Type 2 (RXB2)
+
Satisfactory performance as Division Supervisor (DIVS)
+
Successful position performance as a Strategic Operational Planner (SOPL) on a wildfire

310-1 PHYSICAL FITNESS LEVEL
Moderate

310-1 OTHER POSITION ASSIGNMENTS THAT WILL MAINTAIN CURRENCY
Prescribed Fire Burn Boss Type 2 (RXB2)
Division Supervisor (DIVS)
Incident Commander Type 3 (ICT3)

310-1 OTHER TRAINING WHICH SUPPORTS DEVELOPMENT OF KNOWLEDGE AND SKILLS
M-581 Fire Program Management

Task Book available at:
http://www.nwcg.gov/pms/taskbook/command/command.htm

27.1 - Exhibit 01--Continued

POSITION QUALIFICATIONS SECTIONS
Planning Positions

FIRE BEHAVIOR ANALYST (FBAN)
Position Category: Wildland Fire

310-1 REQUIRED TRAINING
S-490 Advanced Wildland Fire Behavior Calculations
S-590 Advanced Fire Behavior Interpretation
RT-130 Annual Fireline Safety Refresher

FOREST SERVICE ADDITIONAL REQUIRED TRAINING
IS-700 National Incident Management System (NIMS), An Introduction
IS-800 National Response Plan (NRP), An Introduction

310-1 REQUIRED EXPERIENCE
Satisfactory performance as a Division/Group Supervisor (DIVS)
    +
Successful position performance as a Fire Behavior Analyst (FBAN) on a wildland fire incident

310-1 PHYSICAL FITNESS LEVEL
Moderate

310-1 OTHER POSITION ASSIGNMENTS THAT WILL MAINTAIN CURRENCY
None

310-1 OTHER TRAINING WHICH SUPPORTS DEVELOPMENT OF KNOWLEDGE AND SKILLS
S-491 Intermediate National Fire Danger Rating System

Task Book available at:
http://www.nwcg.gov/pms/taskbook/planning/planning.htm

27.1 - Exhibit 01--Continued

POSITION QUALIFICATIONS SECTIONS
Planning Positions

FIELD OBSERVER (FOBS)
Position Category: Wildland Fire

310-1 REQUIRED TRAINING
RT-130 Annual Fireline Safety Refresher

FOREST SERVICE ADDITIONAL REQUIRED TRAINING
IS-700 National Incident Management System (NIMS), An Introduction

310-1 REQUIRED EXPERIENCE
Satisfactory performance as any single resource boss (CRWB, ENGB, TRPB, FIRB, FELB, HMGB, HEQB)
+
Successful position performance as a Field Observer (FOBS) on a wildland fire incident

310-1 PHYSICAL FITNESS LEVEL
Moderate

310-1 OTHER POSITION ASSIGNMENTS THAT WILL MAINTAIN CURRENCY
Any single resource boss (CRWB, ENGB, TRPB, FIRB, FELB, HMGB, HEQB)
Fire Effects Monitor (FEMO)

310-1 OTHER TRAINING WHICH SUPPORTS DEVELOPMENT OF KNOWLEDGE AND SKILLS
S-244 Field Observer

Task Book available at:
http://www.nwcg.gov/pms/taskbook/planning/planning.htm

27.1 - Exhibit 01--Continued

POSITION QUALIFICATIONS SECTIONS
Planning Positions

TRAINING SPECIALIST (TNSP)
Position Category: Incident Support

310-1 REQUIRED TRAINING
None

FOREST SERVICE ADDITIONAL REQUIRED TRAINING
I-100 Introduction to Incident Command System*
IS-700 National Incident Management System (NIMS), An Introduction
S-445 Incident Training Specialist

310-1 REQUIRED EXPERIENCE
Successful position performance as a Training Specialist (TNSP)

310-1 PHYSICAL FITNESS LEVEL
None Required

310-1 OTHER POSITION ASSIGNMENTS THAT WILL MAINTAIN CURRENCY
None

310-1 OTHER TRAINING WHICH SUPPORTS DEVELOPMENT OF KNOWLEDGE AND SKILLS
I-300 Intermediate Incident Command System
L-180 Human Factors on the Fireline
S-110 Basic Wildland Fire Orientation
S-445 Incident Training Specialist

Task Book available at:
http://www.nwcg.gov/pms/taskbook/planning/planning.htm
*I-100 Online link: *http://training.nwcg.gov/classes/i100.htm*

27.1 - Exhibit 01--Continued

POSITION QUALIFICATIONS SECTIONS
Planning Positions

HUMAN RESOURCE SPECIALIST (HRSP)
Position Category: Incident Support

310-1 REQUIRED TRAINING
S-340 Human Resource Specialist
RT-340 Human Resource Specialist Refresher Workshop (triennial requirement after certification in the position occurs)

FOREST SERVICE ADDITIONAL REQUIRED TRAINING:
I-100 Introduction to Incident Command System*
IS-700 National Incident Management System (NIMS), An Introduction
S-260 Interagency Incident Business Management

310-1 REQUIRED EXPERIENCE
Successful position performance as a Human Resource Specialist (HRSP)

310-1 PHYSICAL FITNESS LEVEL
None Required

310-1 OTHER POSITION ASSIGNMENTS THAT WILL MAINTAIN CURRENCY
None

310-1 OTHER TRAINING WHICH SUPPORTS DEVELOPMENT OF KNOWLEDGE AND SKILLS
L-180 Human Factors on the Fireline
S-110 Basic Wildland Fire Orientation

Task Book available at:
http://www.nwcg.gov/pms/taskbook/planning/planning.htm
*I-100 Online link: *http://training.nwcg.gov/classes/i100.htm*

27.1 - Exhibit 01--Continued

POSITION QUALIFICATIONS SECTIONS
Planning Positions

DISPLAY PROCESSOR (DPRO)
Position Category: Incident Support

310-1 REQUIRED TRAINING
None

FOREST SERVICE ADDITIONAL REQUIRED TRAINING
I-100 Introduction to Incident Command System*
IS-700 National Incident Management System (NIMS), An Introduction

310-1 REQUIRED EXPERIENCE
Successful position performance as a Display Processor (DPRO)

310-1 PHYSICAL FITNESS LEVEL
None Required

310-1 OTHER POSITION ASSIGNMENTS THAT WILL MAINTAIN CURRENCY
Situation Unit Leader (SITL)

310-1 OTHER TRAINING WHICH SUPPORTS DEVELOPMENT OF KNOWLEDGE AND SKILLS
S-110 Basic Wildland Fire Orientation
S-245 Display Processor

Task Book available at:
http://www.nwcg.gov/pms/taskbook/planning/planning.htm
*I-100 Online link: http://training.nwcg.gov/classes/i100.htm

POSITION QUALIFICATIONS SECTIONS
Planning Positions

LONG TERM FIRE ANALYST (LTAN)
Position Category: Wildland Fire

310-1 REQUIRED TRAINING
RT-130 Annual Wildland Fire Safety Refresher (required annually after initial training)
S-390 Introduction to Wildland Fire Behavior Calculations
S-490 Advanced Wildland Fire Behavior Calculations
S-491 Intermediate Fire Danger Rating System
S-495 Geospatial Fire Analysis, Interpretation, And Application
S-590 Advanced Fire Behavior Interpretation

FOREST SERVICE ADDITIONAL REQUIRED TRAINING
IS-700 National Incident Management System (NIMS), An Introduction
S-491 National Fire Danger Rating System

310-1 REQUIRED EXPERIENCE
Satisfactory performance as a Fire Behavior Analyst (FBAN)
        +
Successful performance as a Long Term Fire Analyst (LTAN) on a wildland fire incident
        OR
Satisfactory performance as a Fire Effects Monitor (FEMO)
        +
Satisfactory performance as a Firing Boss, Single Resource (FIRB)
        +
Successful position performance as a Long Term Fire Analyst (LTAN) on a wildland fire incident

310-1 PHYSICAL FITNESS LEVEL
Moderate

310-1 OTHER POSITION ASSIGNMENTS THAT WILL MAINTAIN CURRENCY
Fire Behavior Analyst (FBAN)

310-1 OTHER TRAINING WHICH SUPPORTS DEVELOPMENT OF KNOWLEDGE AND SKILLS
RX-410 Smoke Management Techniques
S-482 Advanced Fire Managment
S-491 National Fire Danger Rating System
S-580 Advanced Fire Use Applications
BEHAVE PLUS

Task Book available at:

http://www.nwcg.gov/pms/taskbook/planning/pms-311-29.pdf

POSITION QUALIFICATIONS SECTIONS
Planning Positions

FIRE EFFECTS MONITOR (FEMO)
Position Category: Wildland Fire

310-1 REQUIRED TRAINING
S-290 Intermediate Wildland Fire Behavior
RT-130 Annual Fireline Safety Refresher

FOREST SERVICE ADDITIONAL REQUIRED TRAINING
IS-700 National Incident Management System (NIMS), An Introduction

310-1 REQUIRED EXPERIENCE
Satisfactory performance as a Firefighter Type 2 (FFT2)
        +
Successful position performance as a Fire Effects Monitor (FEMO) on a wildland fire incident

310-1 PHYSICAL FITNESS LEVEL
Moderate

310-1 OTHER POSITION ASSIGNMENTS THAT WILL MAINTAIN CURRENCY
Field Observer (FOBS)
Any higher position for which this position is a prerequisite

310-1 OTHER TRAINING WHICH SUPPORTS DEVELOPMENT OF KNOWLEDGE AND SKILLS
RX-310 Introduction to Fire Effects
S-244 Field Observer

Task Book available at:
http://www.nwcg.gov/pms/taskbook/planning/pms-311-30.pdf

27.1 - Exhibit 01--Continued

POSITION QUALIFICATIONS SECTIONS
Planning Positions

GEOGRAPHIC INFORMATION SYSTEM SPECIALIST (GISS)
Position Category: Incident Support

310-1 REQUIRED TRAINING
None

FOREST SERVICE ADDITIONAL REQUIRED TRAINING

IS-700 National Incident Management System (NIMS), An Introduction
S-341 Geographic Information System (GIS) Specialist for Incident Management

310-1 REQUIRED EXPERIENCE
Successful position performance as a Geographic Information System Specialist (GISS)

310-1 PHYSICAL FITNESS
None Required

310-1 OTHER POSITION ASSIGNMENTS THAT WILL MAINTAIN CURRENCY
None

310-1 OTHER TRAINING WHICH SUPPORTS DEVELOPMENT OF KNOWLEDGE AND SKILLS
I-100 Introduction to Incident Command System
S-110 Basic Wildland Fire Orientation
S-130 Firefighter Training
S-190 Introduction to Wildland Fire Behavior
S-245 Display Processor
S-495 Geospatial Fire Analysis, Interpretation, and Application

Task Book available at:
http://www.nwcg.gov/pms/taskbook/planning/planning.htm

27.1 - Exhibit 01--Continued

POSITION QUALIFICATIONS SECTIONS
Logistics Positions

ASSISTANT AREA COMMANDER, LOGISTICS (ACLC)
Position Category: Incident Command System

310-1 REQUIRED TRAINING
S-620 Area Command

FOREST SERVICE ADDITIONAL REQUIRED TRAINING
IS-700 National Incident Management System (NIMS), An Introduction
IS-800 National Response Plan (NRP), An Introduction

310-1 REQUIRED EXPERIENCE
Satisfactory performance as an Incident Commander or General Staff on a National Type 1
Incident Management Team
     +
Successful position performance as an Assistant Area Commander, Logistics (ACLC)

310-1 PHYSICAL FITNESS LEVEL
None Required

310-1 OTHER POSITION ASSIGNMENTS THAT WILL MAINTAIN CURRENCY
Assistant Area Commander, Planning (ACPC)
Incident Commander Type 1 (ICT1)
Any higher position for which this position is a prerequisite

Task Book available at:
http://www.nwcg.gov/pms/taskbook/logistics/logistic.htm

27.1 - Exhibit 01--Continued

POSITION QUALIFICATIONS SECTIONS
Logistics Positions

LOGISTICS SECTION CHIEF TYPE I (LSC1)
Position Category: Incident Command System

310-1 REQUIRED TRAINING
S-520 Advanced Incident Management

FOREST SERVICE ADDITIONAL REQUIRED TRAINING
IS-700 National Incident Management System (NIMS), An Introduction
IS-800 National Response Plan (NRP), An Introduction

310-1 REQUIRED EXPERIENCE
Satisfactory performance as a Logistics Section Chief Type 2 (LSC2)
         +
Successful position performance as a Logistics Section Chief Type 1 (LSC1)

310-1 PHYSICAL FITNESS LEVEL
None Required

310-1 OTHER POSITION ASSIGNMENTS THAT WILL MAINTAIN CURRENCY
Logistics Section Chief Type 2 (LSC2)
Service Branch Director (SVBD)
Support Branch Director (SUBD)
Any higher position for which this position is a prerequisite

310-1 OTHER TRAINING WHICH SUPPORTS DEVELOPMENT OF KNOWLEDGE AND SKILLS
None

Task Book available at:
http://www.nwcg.gov/pms/taskbook/logistics/logistic.htm

27.1 - Exhibit 01--Continued

POSITION QUALIFICATIONS SECTIONS
Logistics Positions

LOGISTICS SECTION CHIEF TYPE 2 (LSC2)
Position Category: Incident Command System

310-1 REQUIRED TRAINING
S-420 Command and General Staff

FOREST SERVICE ADDITIONAL REQUIRED TRAINING
IS-700 National Incident Management System (NIMS), An Introduction
IS-800 National Response Plan (NRP), An Introduction

310-1 REQUIRED EXPERIENCE
Satisfactory performance as a Facilities Unit Leader (FACL)
+
Satisfactory performance as a Ground Support Unit Leader (GSUL)
+
Successful position performance as a Logistics Section Chief Type 2 (LSC2)
OR
Satisfactory performance as a Facilities Unit Leader (FACL)
+
Satisfactory performance as a Supply Unit Leader (SPUL)
+
Successful position performance as a Logistics Section Chief Type 2 (LSC2)

310-1 PHYSICAL FITNESS LEVEL
None Required

310-1 OTHER POSITION ASSIGNMENTS THAT WILL MAINTAIN CURRENCY
Service Branch Director (SVBD)
Support Branch Director (SUBD)
Facilities Unit Leader (FACL)
Ground Support Unit Leader (GSUL)
Supply Unit Leader (SPUL)
Any higher position for which this position is a prerequisite

27.1 - Exhibit 01--Continued

POSITION QUALIFICATIONS SECTIONS
Logistics Positions

LOGISTICS SECTION CHIEF TYPE 2 (LSC2) - CONTINUED
Position Category: Incident Command System

310-1 OTHER TRAINING WHICH SUPPORTS DEVELOPMENT OF KNOWLEDGE AND SKILLS
I-400 Advanced Incident Command System
L-480 Organizational Leadership in the Fire Service
S-450 Logistics Section Chief

Task Book available at:
http://www.nwcg.gov/pms/taskbook/logistics/logistic.htm

POSITION QUALIFICATIONS SECTIONS
Logistics Positions

SERVICE BRANCH DIRECTOR (SVBD)
Position Category: Incident Command System

310-1 REQUIRED TRAINING
None

310-1 REQUIRED EXPERIENCE
Satisfactory performance as a Logistics Section Chief Type 2 (LSC2)

FOREST SERVICE ADDITIONAL REQUIRED TRAINING
IS-700 National Incident Management System (NIMS), An Introduction
IS-800 National Response Plan (NRP), An Introduction

310-1 PHYSICAL FITNESS LEVEL
None Required

310-1 OTHER POSITION ASSIGNMENTS THAT WILL MAINTAIN CURRENCY
Support Branch Director (SUBD)
Logistics Section Chief Type 2 (LSC2)
Medical Unit Leader (MEDL)
Communications Unit Leader (COML)
Food Unit Leader (FDUL)

310-1 OTHER TRAINING WHICH SUPPORTS DEVELOPMENT OF KNOWLEDGE AND SKILLS
None

27.1 - Exhibit 01--Continued

POSITION QUALIFICATIONS SECTIONS
Logistics Positions

SUPPORT BRANCH DIRECTOR (SUBD)
Position Category: Incident Command System

310-1 REQUIRED TRAINING
None

FOREST SERVICE ADDITIONAL REQUIRED TRAINING
IS-700 National Incident Management System (NIMS), An Introduction
IS-800 National Response Plan (NRP), An Introduction

310-1 REQUIRED EXPERIENCE
Satisfactory performance as a Logistics Section Chief Type 2 (LSC2)

310-1 OTHER POSITION ASSIGNMENTS THAT WILL MAINTAIN CURRENCY
Logistics Section Chief Type 2 (LSC2)
Service Branch Director (SVBD)
Facilities Unit Leader (FACL)
Ground Support Unit Leader (GSUL)
Supply Unit Leader (SUPL)

310-1 OTHER TRAINING WHICH SUPPORTS DEVELOPMENT OF KNOWLEDGE AND SKILLS
None

POSITION QUALIFICATIONS SECTIONS
Logistics Positions

MEDICAL UNIT LEADER (MEDL)
Position Category: Incident Command System

310-1 REQUIRED TRAINING
S-359 Medical Unit Leader

FOREST SERVICE ADDITIONAL REQUIRED TRAINING
I-100 Introduction to Incident Command System*
IS-700 National Incident Management System (NIMS), An Introduction
IS-800 National Response Plan (NRP), An Introduction

FOREST SERVICE ANNUAL CERTIFICATION REQUIREMENT
Emergency Medical Technician (any)

FOREST SERVICE ANNUAL LICENSE REQUIREMENT
State License

310-1 REQUIRED EXPERIENCE
Prior or current certification as an Emergency Medical Technician (EMT) or equivalent
       +
Successful position performance as a Medical Unit Leader (MEDL)

310-1 PHYSICAL FITNESS LEVEL
None Required

310-1 OTHER POSITION ASSIGNMENTS THAT WILL MAINTAIN CURRENCY
None

310-1 OTHER TRAINING WHICH SUPPORTS DEVELOPMENT OF KNOWLEDGE AND SKILLS
I-300 Intermediate Incident Command System
L-380 Fireline Leadership
S-110 Basic Wildland Fire Orientation

Task Book available at:
http://www.nwcg.gov/pms/taskbook/logistics/logistic.htm
*I-100 Online link: *http://training.nwcg.gov/classes/i100.htm*

27.1 - Exhibit 01--Continued

POSITION QUALIFICATIONS SECTIONS
Logistics Positions

COMMUNICATIONS UNIT LEADER (COML)
Position Category: Incident Command System

310-1 REQUIRED TRAINING
None

FOREST SERVICE ADDITIONAL REQUIRED TRAINING
IS-700 National Incident Management System (NIMS), An Introduction
IS-800 National Response Plan (NRP), An Introduction

310-1 REQUIRED EXPERIENCE
Satisfactory performance as an Incident Communications Technician (COMT)
    +
Satisfactory performance as an Incident Communications Center Manager (INCM)
    +
Successful position performance as a Communications Unit Leader (COML)

310-1 PHYSICAL FITNESS LEVEL
None Required

310-1 OTHER POSITION ASSIGNMENTS THAT WILL MAINTAIN CURRENCY
Incident Communications Center Manager (INCM)
Incident Communications Technician (COMT)

310-1 OTHER TRAINING WHICH SUPPORTS DEVELOPMENT OF KNOWLEDGE AND SKILLS
I-300 Intermediate Incident Command System
L-380 Fireline Leadership
S-358 Communications Unit Leader

Task Book available at:
http://www.nwcg.gov/pms/taskbook/logistics/logistic.htm

27.1 - Exhibit 01--Continued

POSITION QUALIFICATIONS SECTIONS
Logistics Positions

INCIDENT COMMUNICATIONS TECHNICIAN (COMT)
Position Category: Incident Command System

310-1 REQUIRED TRAINING
RT-130 Annual Fireline Safety Refresher

FOREST SERVICE ADDITIONAL REQUIRED TRAINING
I-100 Introduction to Incident Command System*
IS-700 National Incident Management System (NIMS), An Introduction

310-1 REQUIRED EXPERIENCE
Successful position performance as an Incident Communications Technician (COMT)

310-1 PHYSICAL FITNESS LEVEL
Light

310-1 OTHER POSITION ASSIGNMENTS THAT WILL MAINTAIN CURRENCY
Any higher position for which this position is a prerequisite

310-1 OTHER TRAINING WHICH SUPPORTS DEVELOPMENT OF KNOWLEDGE AND SKILLS
Basic Firefighter Training:
I-100 Introduction to Incident Command System
L-180 Human Factors on the Fireline
S-130 Firefighter Training
S-190 Introduction to Wildland Fire Behavior
S-258 Incident Communications Technician

Task Book available at:
http://www.nwcg.gov/pms/taskbook/logistics/logistic.htm
*I-100 Online link: *http://training.nwcg.gov/classes/i100.htm*

27.1 - Exhibit 01--Continued

POSITION QUALIFICATIONS SECTIONS
Logistics Positions

INCIDENT COMMUNICATIONS CENTER MANAGER (INCM)
Position Category: Incident Command System

310-1 REQUIRED TRAINING
None

FOREST SERVICE ADDITIONAL REQUIRED TRAINING
IS-700 National Incident Management System (NIMS), An Introduction

310-1 REQUIRED EXPERIENCE
Satisfactory performance as a Radio Operator (RADO)
+
Successful position performance as an Incident Communications Center Manager (INCM)

310-1 PHYSICAL FITNESS LEVEL
None Required

310-1 OTHER POSITION ASSIGNMENTS THAT WILL MAINTAIN CURRENCY
Any higher position for which this position is a prerequisite

310-1 OTHER TRAINING WHICH SUPPORTS DEVELOPMENT OF KNOWLEDGE AND SKILLS
I-200 Basic Incident Command System
J-257 Incident Communications Center Manager
S-110 Basic Wildland Fire Orientation
S-260 Interagency Incident Business Management

Task Book available at:
http://www.nwcg.gov/pms/taskbook/logistics/logistic.htm

27.1 - Exhibit 01--Continued

POSITION QUALIFICATIONS SECTIONS
Logistics Positions

RADIO OPERATOR (RADO)
Position Category: Incident Support

310-1 REQUIRED TRAINING
None

FOREST SERVICE ADDITIONAL REQUIRED TRAINING
I-100 Introduction to Incident Command System*
IS-700 National Incident Management System (NIMS), An Introduction

310-1 REQUIRED EXPERIENCE
Successful position performance as a Radio Operator (RADO)

310-1 PHYSICAL FITNESS LEVEL
None Required

310-1 OTHER POSITION ASSIGNMENTS THAT WILL MAINTAIN CURRENCY
Aircraft Base Radio Operator (ABRO)
Initial Attack Dispatcher (IADP)
Any higher position for which this position is a prerequisite

310-1 OTHER TRAINING WHICH SUPPORTS DEVELOPMENT OF KNOWLEDGE AND SKILLS
Basic Firefighter Training:
I-100 Introduction to Incident Command System
L-180 Human Factors on the Fireline
S-130 Firefighter Training
S-190 Introduction to Wildland Fire Behavior
J-158 Radio Operator

Task Book available at:
http://www.nwcg.gov/pms/taskbook/logistics/logistic.htm
*I-100 Online link: *http://training.nwcg.gov/classes/i100.htm*

27.1 - Exhibit 01--Continued

POSITION QUALIFICATIONS SECTIONS
Logistics Positions

FOOD UNIT LEADER (FDUL)
Position Category: Incident Command System

310-1 REQUIRED TRAINING
S-357 Food Unit Leader

FOREST SERVICE ADDITIONAL REQUIRED TRAINING
I-100 Introduction to Incident Command System*
IS-700 National Incident Management System (NIMS), An Introduction
IS-800 National Response Plan (NRP), An Introduction

310-1 REQUIRED EXPERIENCE
Successful position performance as a Food Unit Leader (FDUL)

310-1 PHYSICAL FITNESS LEVEL
None Required

310-1 OTHER POSITION ASSIGNMENTS THAT WILL MAINTAIN CURRENCY
None

310-1 OTHER TRAINING WHICH SUPPORTS DEVELOPMENT OF KNOWLEDGE AND SKILLS
I-300 Intermediate Incident Command System
L-380 Fireline Leadership
S-110 Basic Wildland Fire Orientation

Task Book available at:
http://www.nwcg.gov/pms/taskbook/logistics/logistic.htm
*I-100 Online link: *http://training.nwcg.gov/classes/i100.htm*

27.1 - Exhibit 01--Continued

POSITION QUALIFICATIONS SECTIONS
Logistics Positions

SUPPLY UNIT LEADER (SPUL)
Position Category: Incident Command System

310-1 REQUIRED TRAINING
None

FOREST SERVICE ADDITIONAL REQUIRED TRAINING
IS-700 National Incident Management System (NIMS), An Introduction
IS-800 National Response Plan (NRP), An Introduction

310-1 REQUIRED EXPERIENCE
Satisfactory performance as an Ordering Manager (ORDM)
    +
Satisfactory performance as a Receiving/Distribution Manager (RCDM)
    +
Successful position performance as a Supply Unit Leader (SPUL)

310-1 PHYSICAL FITNESS LEVEL
None Required

310-1 OTHER POSITION ASSIGNMENTS THAT WILL MAINTAIN CURRENCY
Ordering Manager (ORDM)
Receiving/Distribution Manager (RCDM)
Any higher position for which this position is a prerequisite

310-1 OTHER TRANIING WHICH SUPPORTS DEVELOPMENT OF KNOWLEDGE AND SKILLS
I-300 Intermediate Incident Command System
L-380 Fireline Leadership
S-356 Supply Unit Leader

Task Book available at:
http://www.nwcg.gov/pms/taskbook/logistics/logistic.htm

27.1 - Exhibit 01--Continued

POSITION QUALIFICATIONS SECTIONS
Logistics Positions

ORDERING MANAGER (ORDM)
Position Category: Incident Command System

310-1 REQUIRED TRAINING
None

FOREST SERVICE ADDITIONAL REQUIRED TRAINING
IS-700 National Incident Management System (NIMS), An Introduction

310-1 REQUIRED EXPERIENCE

Successful position performance as an Ordering Manager (ORDM)

310-1 PHYSICAL FITNESS LEVEL
None Required

310-1 OTHER POSITION ASSIGNMENTS THAT WILL MAINTAIN CURRENCY
Receiving/Distribution Manager (RCDM)
Expanded Dispatch Recorder (EDRC)
Base/Camp Manager (BCMG)
Equipment Manager (EQPM)
Any higher position for which this position is a prerequisite

310-1 OTHER TRAINING WHICH SUPPORTS DEVELOPMENT OF KNOWLEDGE AND SKILLS
I-200 Basic Incident Command System
J-252 Ordering Manager
S-110 Basic Wildland Fire Orientation
S-260 Interagency Incident Business Management

Task Book available at:
http://www.nwcg.gov/pms/taskbook/logistics/logistic.htm

27.1 - Exhibit 01--Continued

POSITION QUALIFICATIONS SECTIONS
Logistics Positions

RECEIVING/DISTRIBUTION MANAGER (RCDM)
Position Category: Incident Command System

310-1 REQUIRED TRAINING
None

FOREST SERVICE ADDITIONAL REQUIRED TRAINING
I-100 Introduction to Incident Command System*
IS-700 National Incident Management System (NIMS), An Introduction

310-1 REQUIRED EXPERIENCE
Successful position performance as a Receiving/Distribution Manager (RCDM)

310-1 PHYSICAL FITNESS LEVEL
None Required

310-1 OTHER POSITION ASSIGNMENTS THAT WILL MAINTAIN CURRENCY
Ordering Manager (ORDM)
Any higher position for which this position is a prerequisite

310-1 OTHER TRAINING WHICH SUPPORTS DEVELOPMENT OF KNOWLEDGE AND SKILLS
I-200 Basic Incident Command System
J-253 Receiving and Distribution Manager
L-180 Human Factors on the Fireline
S-110 Basic Wildland Fire Orientation
S-260 Interagency Incident Business Management

Task Book available at:
http://www.nwcg.gov/pms/taskbook/logistics/logistic.htm
*I-100 Online link: *http://training.nwcg.gov/classes/i100.htm*

<u>**27.1 - Exhibit 01--Continued**</u>

POSITION QUALIFICATIONS SECTIONS
Logistics Positions

FACILITIES UNIT LEADER (FACL)
Position Category: Incident Command System

310-1 REQUIRED TRAINING
None

FOREST SERVICE ADDITIONAL REQUIRED TRAINING
IS-700 National Incident Management System (NIMS), An Introduction
IS-800 National Response Plan (NRP), An Introduction

310-1 REQUIRED EXPERIENCE
Satisfactory performance as a Base Camp Manager (BCMG)
    +
Successful position performance as a Facilities Unit Leader (FACL)

310-1 PHYSICAL FITNESS LEVEL
None Required

310-1 OTHER POSITION ASSIGNMENTS THAT WILL MAINTAIN CURRENCY
Ground Support Unit Leader (GSUL)
Supply Unit Leader (SPUL)
Base/Camp Manager (BCMG)
Equipment Manager (EQPM)
Ordering Manager (ORDM)
Receiving/Distribution Manager (RCDM)
Security Manager (SECM)
Any higher position for which this position is a prerequisite

310-1 OTHER TRAINING WHICH SUPPORTS DEVELOPMENT OF KNOWLLEDGE AND SKILLS
I-300 Intermediate Incident Command System
L-380 Fireline Leadership
S-354 Facilities Unit Leader

Task Book available at:
http://www.nwcg.gov/pms/taskbook/logistics/logistic.htm

27.1 - Exhibit 01--Continued

POSITION QUALIFICATIONS SECTIONS
Logistics Positions

SECURITY MANAGER (SECM)
Position Category: Incident Command System

310-1 REQUIRED TRAINING
None

FOREST SERVICE ADDITIONAL REQUIRED TRAINING
I-100 Introduction to Incident Command System*
IS-700 National Incident Management System (NIMS), An Introduction

310-1 REQUIRED EXPERIENCE
Successful position performance as a Security Manager (SECM)

FOREST SERVICE REQUIRED EXPERIENCE
Successful completion of a basic land management police training program or a criminal investigator training program conducted by the Federal Law Enforcement Training Center**

310-1 PHYSICAL FITNESS LEVEL
None

310-1 OTHER POSITION ASSIGNMENTS THAT WILL MAINTAIN CURRENCY
None

310-1 OTHER TRAINING WHICH SUPPORTS DEVELOPMENT OF KNOWLEDGE AND SKILLS
I-200 Basic Incident Command System
J-259 Security Manager
L-180 Human Factors on the Fireline
S-110 Basic Wildland Fire Orientation
S-260 Interagency Incident Business Management

Task Book available at:
http://www.nwcg.gov/pms/taskbook/logistics/logistic.htm
*I-100 Online link: *http://training.nwcg.gov/classes/i100.htm*
**Certification of a Security Manager who has met this required experience does not include completion of a position task book. Competency for the position task book can be granted by indicating the year of completion of the training programs provided.

27.1 - Exhibit 01--Continued

POSITION QUALIFICATIONS SECTIONS
Logistics Positions

BASE CAMP MANAGER (BCMG)
Position Category: Incident Command System

310-1 REQUIRED TRAINING
None

FOREST SERVICE ADDITIONAL REQUIRED TRAINING
I-100 Introduction to Incident Command System*
IS-700 National Incident Management System (NIMS), An Introduction

310-1 REQUIRED EXPERIENCE
Successful position performance as a Base/Camp Manager (BCMG)

310-1 PHYSICAL FITNESS LEVEL
Light

310-1 OTHER POSITION ASSIGNMENTS THAT WILL MAINTAIN CURRENCY
Equipment Manager (EQPM)
Ordering Manager (ORDM)
Receiving/Distribution Manager (RCDM)
Any higher position for which this position is a prerequisite

310-1 OTHER TRAINING WHICH SUPPORTS DEVELOPMENT OF KNOWLEDGE AND SKILLS
I-200 Basic Incident Command System
J-254 Base/Camp Manager
L-180 Human Factors on the Fireline
S-110 Basic Wildland Fire Orientation
S-260 Interagency Incident Business Management

Task Book available at:
http://www.nwcg.gov/pms/taskbook/logistics/logistic.htm
*I-100 Online link: *http://training.nwcg.gov/classes/i100.htm*

27.1 - Exhibit 01--Continued

POSITION QUALIFICATIONS SECTIONS
Logistics Positions

GROUND SUPPORT UNIT LEADER (GSUL)
Position Category: Incident Command System

310-1 REQUIRED TRAINING
None

FOREST SERVICE ADDITIONAL REQUIRED TRAINING
IS-700 National Incident Management System (NIMS), An Introduction
IS-800 National Response Plan (NRP), An Introduction

310-1 REQUIRED EXPERIENCE
Satisfactory performance as an Equipment Manager (EQPM)
       +
Successful position performance as a Ground Support Unit Leader (GSUL)

310-1 PHYSICAL FITNESS LEVEL
None Required

310-1 OTHER POSITION ASSIGNMENTS THAT WILL MAINTAIN CURRENCY
Facilities Unit Leader (FACL)
Supply Unit Leader (SPUL)
Equipment Manager (EQPM)
Base/Camp Manager (BCMG)
Ordering Manager (ORDM)
Receiving/Distribution Manager (RCDM)
Any higher position for which this position is a prerequisite

310-1 OTHER TRAINING WHICH SUPPORTS DEVELOPMENT OF KNOWLEDGE AND SKILLS
I-300 Intermediate Incident Command System
L-380 Fireline Leadership
S-355 Ground Support Unit Leader

Task Book available at:
http://www.nwcg.gov/pms/taskbook/logistics/logistic.htm

27.1 - Exhibit 01--Continued

POSITION QUALIFICATIONS SECTIONS
Logistics Positions

EQUIPMENT MANAGER (EQPM)
Position Category: Incident Command System

310-1 REQUIRED TRAINING
None

FOREST SERVICE ADDITIONAL REQUIRED TRAINING
I-100 Introduction to Incident Command System*
IS-700 National Incident Management System (NIMS), An Introduction

310-1 REQUIRED EXPERIENCE
Successful position performance as an Equipment Manager (EQPM)

310-1 PHYSICAL FITNESS LEVEL
None Required

310-1 OTHER POSITION ASSIGNMENTS THAT WILL MAINTAIN CURRENCY
Base/Camp Manager (BCMG)
Ordering Manager (ORDM)
Receiving/Distribution Manager (RCDM)
Any higher position for which this position is a prerequisite

310-1 OTHER TRAINING WHICH SUPPORTS DEVELOPMENT OF KNOWLEDGE AND SKILLS
I-200 Basic Incident Command System
J-255 Equipment Manager
L-180 Human Factors on the Fireline
S-110 Basic Wildland Fire Orientation
S-260 Interagency Incident Business Management

Task Book available at:
http://www.nwcg.gov/pms/taskbook/logistics/logistic.htm
*I-100 Online link: *http://training.nwcg.gov/classes/i100.htm*

27.1 - Exhibit 01--Continued

POSITION QUALIFICATIONS SECTIONS
Finance/Administration Positions

FINANCE/ADMINISTRATION SECTION CHIEF TYPE 1 (FSC1)
Position Category: Incident Command System

310-1 REQUIRED TRAINING
S-520 Advanced Incident Management

FOREST SERVICE ADDITIONAL REQUIRED TRAINING
IS-700 National Incident Management System (NIMS), An Introduction
IS-800 National Response Plan (NRP), An Introduction

310-1 REQUIRED EXPERIENCE
Satisfactory performance as a Finance/Administration Section Chief Type 2 (FSC2)
    +
Successful position performance as a Finance/Administration Section Chief Type 1 (FSC1)

310-1 PHYSICAL FITNESS LEVEL
None Required

310-1 OTHER POSITION ASSIGNMENTS THAT WILL MAINTAIN CURRENCY
Finance/Administration Section Chief Type 2 (FSC2)
Incident Commander Type 1 (ICT1)
Incident Business Advisor Type 1 (IBA1)
Any higher position for which this position is a prerequisite

310-1 OTHER TRAINING WHICH SUPPORTS DEVELOPMENT OF KNOWLEDGE AND SKILLS
None

Task Book available at:
http://www.nwcg.gov/pms/taskbook/finance/finance.htm

27.1 - Exhibit 01--Continued

POSITION QUALIFICATIONS SECTIONS
Finance/Administration Positions

FINANCE/ADMINISTRATION SECTION CHIEF TYPE 2 (FSC2)
Position Category: Incident Command System

310-1 REQUIRED TRAINING
S-420 Command and General Staff

FOREST SERVICE ADDITIONAL REQUIRED TRAINING
IS-700 National Incident Management System (NIMS), An Introduction
IS-800 National Response Plan (NRP), An Introduction

310-1 REQUIRED EXPERIENCE
Satisfactory performance as a Time Unit Leader (TIME)
      +
Satisfactory performance as a Procurement Unit Leader (PROC)
      +
Successful position performance as a Finance/Administration Section Chief Type 2 (FSC2)
      OR
Satisfactory performance as a Time Unit Leader (TIME)
      +
Satisfactory performance as a Cost Unit Leader (COST)
      +
Successful position performance as a Finance/Administration Section Chief Type 2 (FSC2)

310-1 PHYSICAL FITNESS LEVEL
None Required

310-1 OTHER POSITION ASSIGNMENTS THAT WILL MAINTAIN CURRENCY
Cost Unit Leader (COST)
Procurement Unit Leader (PROC)
Time Unit Leader (Time)
Compensation/Claims Unit Leader (COMP)
Incident Commander Type 2 (ICT2)
Incident Business Advisory Type 2 (IBA2)
Any higher position for which this position is a prerequisite

27.1 - Exhibit 01--Continued

POSITION QUALIFICATIONS SECTIONS
Finance/Administration Positions

FINANCE/ADMINISTRATION SECTION CHIEF TYPE 2 (FSC2) - CONTINUED
Position Category: Incident Command System

310-1 OTHER TRAINING WHICH SUPPORTS DEVELOPMENT OF KNOWLEDGE AND SKILLS
I-400 Advanced Incident Command System
L-480 Organizational Leadership in the Fire Service
S-460 Finance/Administration Section Chief

Task Book available at:
http://www.nwcg.gov/pms/taskbook/finance/finance.htm

27.1 - Exhibit 01--Continued

POSITION QUALIFICATIONS SECTIONS
Finance/Administration Positions

TIME UNIT LEADER (TIME)
Position Category: Incident Command System

310-1 REQUIRED TRAINING
None

FOREST SERVICE ADDITIONAL REQUIRED TRAINING
IS-700 National Incident Management System (NIMS), An Introduction
IS-800 National Response Plan (NRP), An Introduction
S-360 Finance/Administrative Unit Leader

310-1 REQUIRED EXPERIENCE
Satisfactory performance as a Personnel Time Recorder (PTRC)
    +
Successful position performance as a Time Unit Leader (TIME)

310-1 PHYSICAL FITNESS LEVEL
None Required

310-1 OTHER POSITION ASSIGNMENTS THAT WILL MAINTAIN CURRENCY
Personnel Time Recorder (PTRC)
Equipment Time Recorder (EQTR)
Any higher position for which this position is a prerequisite

310-1 OTHER TRAINING WHICH SUPPORTS DEVELOPMENT OF KNOWLEDGE AND SKILLS
I-300 Intermediate Incident Command System
L-380 Fireline Leadership
S-360 Finance/Administration Unit Leader

Task Book available at:
http://www.nwcg.gov/pms/taskbook/finance/finance.htm

27.1 - Exhibit 01--Continued

POSITION QUALIFICATIONS SECTIONS
Finance/Administration Positions

PERSONNEL TIME RECORDER (PTRC)
Position Category: Incident Command System

310-1 REQUIRED TRAINING
None

FOREST SERVICE ADDITIONAL REQUIRED TRAINING
I-100 Introduction to Incident Command System*

IS-700 National Incident Management System (NIMS), An Introduction
S-260 Interagency Incident Business Management
S-261 Applied Interagency Incident Business Management

310-1 REQUIRED EXPERIENCE
Successful position performance as a Personnel Time Recorder (PTRC)

FOREST SERVICE ADDITIONAL REQUIRED EXPERIENCE
On-the-job experience in Incident Base Automation (I-SUITE)

310-1 PHYSICAL FITNESS LEVEL
None Required

310-1 OTHER POSITION ASSIGNMENTS THAT WILL MAINTAIN CURRENCY
Equipment Time Recorder (EQTR)
Any higher position for which this position is a prerequisite

310-1 OTHER TRAINING WHICH SUPPORTS DEVELOPMENT OF KNOWLEDGE AND SKILLS
I-SUITE Incident Base Automation
I-100 Introduction to Incident Command System
L-180 Human Factors on the Fireline
S-110 Basic Wildland Fire Orientation
S-260 Interagency Incident Business Management
S-261 Applied Interagency Incident Business Management

Task Book available at:
http://www.nwcg.gov/pms/taskbook/finance/finance.htm
*I-100 Online link: *http://training.nwcg.gov/classes/i100.htm*

27.1 - Exhibit 01--Continued

POSITION QUALIFICATIONS SECTIONS
Finance/Administration Positions

COST UNIT LEADER (COST)
Position Category: Incident Command System

310-1 REQUIRED TRAINING
None

FOREST SERVICE ADDITIONAL REQUIRED TRAINING
I-100 Introduction to Incident Command System*
IS-700 National Incident Management System (NIMS), An Introduction
IS-800 National Response Plan (NRP), An Introduction
S-260 Interagency Incident Business Management
S-261 Applied Interagency Incident Business Management
S-360 Finance/Administration Unit Leader
I-SUITE Incident Base Automation

310-1 REQUIRED EXPERIENCE
Successful position performance as a Cost Unit Leader (COST)

310-1 PHYSICAL FITNESS LEVEL
None Required

310-1 OTHER POSITION ASSIGNMENTS THAT WILL MAINTAIN CURRENCY
Any higher position for which this position is a prerequisite

310-1 OTHER TRAINING WHICH SUPPORTS DEVELOPMENT OF KNOWLEDGE AND SKILLS
I-Suite Incident Base Automation
I-300 Intermediate Incident Command System
L-380 Fireline Leadership
S-110 Basic Wildland Fire Orientation
S-260 Interagency Incident Business Management
S-261 Applied Interagency Incident Business Management
S-360 Finance/Administration Unit Leader

Task Book available at:
http://www.nwcg.gov/pms/taskbook/finance/finance.htm
*I-100 Online link: *http://training.nwcg.gov/classes/i100.htm*

27.1 - Exhibit 01--Continued

POSITION QUALIFICATIONS SECTIONS
Finance/Administration Positions

COMMISSARY MANAGER (CMSY)
Position Category: Incident Command System

310-1 REQUIRED TRAINING
None

FOREST SERVICE ADDITIONAL REQUIRED TRAINING
I-100 Introduction to Incident Command System*
IS-700 National Incident Management System (NIMS), An Introduction

310-1 REQUIRED EXPERIENCE
Successful position performance as a Commissary Manager (CMSY)

310-1 PHYSICAL FITNESS LEVEL
None Required

310-1 OTHER POSITION ASSIGNMENTS THAT WILL MAINTAIN CURRENCY
Time Unit Leader (TIME)
Personnel Time Recorder (PTRC)

310-1 OTHER TRAINING WHICH SUPPORTS DEVELOPMENT OF KNOWLEDGE AND SKILLS
I-100 Introduction to Incident Command System
L-180 Human Factors on the Fireline
S-110 Basic Wildland Fire Orientation
S-260 Interagency Incident Business Management
S-261 Applied Interagency Incident Business Management

Task Book available at:
http://www.nwcg.gov/pms/taskbook/finance/finance.htm
*I-100 Online link: *http://training.nwcg.gov/classes/i100.htm*

27.1 - Exhibit 01--Continued

POSITION QUALIFICATIONS SECTIONS
Finance/Administration Positions

COMPENSATION/CLAIMS UNIT LEADER (COMP)
Position Category: Incident Command System

310-1 REQUIRED TRAINING
None

FOREST SERVICE ADDITIONAL REQUIRED TRAINING
IS-700 National Incident Management System (NIMS), An Introduction
IS-800 National Response Plan (NRP), An Introduction

310-1 REQUIRED EXPERIENCE
Satisfactory performance as a Compensation-for-Injury Specialist (INJR)
+
Satisfactory performance as a Claims Specialist (CLMS)
+
Successful position performance as a Compensation/Claims Unit Leader (COMP)

310-1 PHYSICAL FITNESS LEVEL
None Required

310-1 OTHER POSITION ASSIGNMENTS THAT WILL MAINTAIN CURRENCY
Claims Specialist (CLMS)
Compensation-for-Injury Specialist (INJR)
Finance/Administration Section Chief Type 2 (FSC2)

310-1 OTHER TRAINING WHICH SUPPORTS DEVELOPMENT OF KNOWLEDGE AND SKILLS
I-300 Intermediate Incident Command System
L-380 Fireline Leadership
S-360 Finance/Administration Unit Leader

Task Book available at:
http://www.nwcg.gov/pms/taskbook/finance/finance.htm

27.1 - Exhibit 01--Continued

POSITION QUALIFICATIONS SECTIONS
Finance/Administration Positions

COMPENSATION-FOR-INJURY SPECIALIST (INJR)
Position Category: Incident Command System

310-1 REQUIRED TRAINING
None

FOREST SERVICE ADDITIONAL REQUIRED TRAINING
I-100 Introduction to Incident Command System*
IS-700 National Incident Management System (NIMS), An Introduction

310-1 REQUIRED EXPERIENCE
Successful position performance as a Compensation-for-Injury Specialist (INJR)

310-1 PHYSICAL FITNESS LEVEL
None Required

310-1 OTHER POSITION ASSIGNMENTS THAT WILL MAINTAIN CURRENCY
Claims Specialist (CLMS)
Compensation/Claims Unit Leader (COMP)
Any higher position for which this position is a prerequisite

310-1 OTHER TRAINING WHICH SUPPORTS DEVELOPMENT OF KNOWLEDGE AND SKILLS
I-100 Introduction to Incident Command System
L-180 Human Factors on the Fireline
S-110 Basic Wildland Fire Orientation
S-260 Interagency Incident Business Management
S-261 Applied Interagency Incident Business Management

Task Book available at:
http://www.nwcg.gov/pms/taskbook/finance/finance.htm
*I-100 Online link: *http://training.nwcg.gov/classes/i100.htm*

27.1 - Exhibit 01--Continued

POSITION QUALIFICATIONS SECTIONS
Finance/Administration Positions

CLAIMS SPECIALIST (CLMS)
Position Category: Incident Command System

310-1 REQUIRED TRAINING
None

FOREST SERVICE ADDITIONAL REQUIRED TRAINING
I-100 Introduction to Incident Command System*
IS-700 National Incident Management System (NIMS), An Introduction

310-1 REQUIRED EXPERIENCE
Successful position performance as a Claims Specialist (CLMS)

310-1 PHYSICAL FITNESS LEVEL
None Required

310-1 OTHER POSITION ASSIGNMENTS THAT WILL MAINTAIN CURRENCY
Compensation-for-Injury Specialist (INJR)
Compensation/Claims Unit Leader (COMP)
Any higher position for which this position is a prerequisite

310-1 OTHER TRAINING WHICH SUPPORTS DEVELOPMENT OF KNOWLEDGE AND SKILLS
I-100 Introduction to Incident Command System
L-180 Human Factors on the Fireline
S-110 Basic Wildland Fire Orientation
S-260 Interagency Incident Business Management
S-261 Applied Interagency Incident Business Management

Task Book available at:
http://www.nwcg.gov/pms/taskbook/finance/finance.htm
*I-100 Online link: *http://training.nwcg.gov/classes/i100.htm*

27.1 - Exhibit 01--Continued

POSITION QUALIFICATIONS SECTIONS
Finance/Administration Positions

PROCUREMENT UNIT LEADER (PROC)
Position Category: Incident Command System

310-1 REQUIRED TRAINING
None

FOREST SERVICE ADDITIONAL REQUIRED TRAINING
I-100 Introduction to Incident Command System*
IS-700 National Incident Management System (NIMS), An Introduction
IS-800 National Response Plan (NRP), An Introduction

AUTHORITY
Federal delegated acquisition authority to obligate Government funds of $100,000 or greater. The Position Task Book may be initiated prior to receiving delegated authority. Certification may not be granted until delegated authority has been obtained.

310-1 REQUIRED EXPERIENCE
Satisfactory performance as an equipment Time Recorder (EQTR)
        +
Successful position performance as a Procurement Unit Leader (PROC)

FOREST SERVICE REQUIRED EXPERIENCE
Successful position performance as a Procurement Unit Leader (PROC)

The Forest Service deviates from the 310-1 by not requiring prerequisite experience of Equipment Time recorder. Forest Service Procurement Unit Leaders must be qualified Contracting Officers (FSH 6309.32) and, by policy, they will not be mobilized into positions other than those that also require delegated acquisition authority (that is Buying Team Leader).

310-1 PHYSICAL FITNESS LEVEL
None Required

310-1 OTHER POSITION ASSIGNMENTS THAT WILL MAINTAIN CURRENCY
Equipment Time Recorder (EQTR)
Personnel Time Recorder (PTRC)
Buying Team Leader (BUYL)
Any higher position for which this position is a prerequisite
*I-100 Online link: _http://training.nwcg.gov/classes/i100.htm_

27.1 - Exhibit 01--Continued

POSITION QUALIFICATIONS SECTIONS
Finance/Administration Positions

PROCUREMENT UNIT LEADER (PROC) - CONTINUED
Position Category: Incident Command System

310-1 OTHER TRAINING WHICH SUPPORTS DEVELOPMENT OF KNOWLEDGE AND SKILLS
I-300 Intermediate Incident Command System
L-380 Fireline Leadership
S-360 Finance/Administration Unit Leader

Task Book available at:
http://www.nwcg.gov/pms/taskbook/finance/finance.htm

27.1 - Exhibit 01--Continued

POSITION QUALIFICATIONS SECTIONS
Finance/Administration Positions

EQUIPMENT TIME RECORDER (EQTR)
Position Category: Incident Command System

310-1 REQUIRED TRAINING
None

FOREST SERVICE ADDITIONAL REQUIRED TRAINING:
I-100 Introduction to Incident Command System*
IS-700 National Incident Management System (NIMS), An Introduction

310-1 REQUIRED EXPERIENCE
Successful position performance as an Equipment Time Recorder (EQTR)

FOREST SERVICE ADDITIONAL REQUIRED EXPERIENCE
On-the-job experience in Incident Base Automation (I-SUITE)

310-1 PHYSICAL FITNESS LEVEL
None Required

310-1 OTHER POSITION ASSIGNMENTS THAT WILL MAINTAIN CURRENCY
Personnel Time Recorder (PTRC)
Any higher position for which this position is a prerequisite

310-1 OTHER TRAINING WHICH SUPPORTS DEVELOPMENT OF KNOWLEDGE AND SKILLS
I-100 Introduction to Incident Command System
L-180 Human Factors on the Fireline
S-110 Basic Wildland Fire Orientation
S-260 Interagency Incident Business Management
S-261 Applied Interagency Incident Business Management

Task Book available at:
http://www.nwcg.gov/pms/taskbook/finance/finance.htm
*I-100 Online link: *http://training.nwcg.gov/classes/i100.htm*

27.1 - Exhibit 01--Continued

POSITION QUALIFICATIONS SECTIONS
Finance/Administration Positions

INCIDENT BUSINESS ADVISOR TYPE 1 (IBA1)
Position Category: Incident Support

310-1 REQUIRED TRAINING
None

FOREST SERVICE ADDITIONAL REQUIRED TRAINING
IS-700 National Incident Management System (NIMS), An Introduction
IS-800 National Response Plan (NRP), An Introduction

310-1 REQUIRED EXPERIENCE
Satisfactory performance as an Incident Business Advisor Type 2 (IBA2)
        +
Successful position performance as an Incident Business Advisory Type 1 (IBA1)

310-1 PHYSICAL FITNESS LEVEL
None Required

310-1 OTHER POSITION ASSIGNMENTS THAT WILL MAINTAIN CURRENCY
Finance/Administration Section Chief Type 1 (FSC1)

310-1 OTHER TRAINING WHICH SUPPORTS DEVELOPMENT OF KNOWLEDGE AND SKILLS
I-400 Advanced Incident Command System

Task Book available at:
http://www.nwcg.gov/pms/taskbook/finance/finance.htm

27.1 - Exhibit 01--Continued

POSITION QUALIFICATIONS SECTIONS
Finance/Administration Positions

INCIDENT BUSINESS ADVISOR TYPE 2 (IBA2)
Position Category: Incident Support

310-1 REQUIRED TRAINING
None

FOREST SERVICE ADDITIONAL REQUIRED TRAINING
IS-700 National Incident Management System (NIMS), An Introduction
IS-800 National Response Plan (NRP), An Introduction
S-481 Incident Business Advisor

310-1 REQUIRED EXPERIENCE
Successful position performance as an Incident Business Advisor Type 2 (IBA2)

310-1 PHYSICAL FITNESS LEVEL
None Required

310-1 OTHER POSITION ASSIGNMENTS THAT WILL MAINTAIN CURRENCY
Finance/Administration Section Chief Type 2 (FSC2)
Any higher position for which this position is a prerequisite

310-1 OTHER TRAINING WHICH SUPPORTS DEVELOPMENT OF KNOWLEDGE AND SKILLS
I-300 Intermediate Incident Command System
S-360 Finance/Administration Unit Leader
S-420 Command and General Staff
S-481 Incident Business Advisor

Task Book available at:
http://www.nwcg.gov/pms/taskbook/finance/finance.htm

POSITION QUALIFICATIONS SECTIONS
Dispatch Positions

EXPANDED DISPATCH COORDINATOR (CORD)
Position Category: Incident Support

310-1 REQUIRED TRAINING
None

FOREST SERVICE ADDITIONAL REQUIRED TRAINING
M-480 Multi-agency Coordinating Group
IS-700 National Incident Management System (NIMS), An Introduction
IS-800 National Response Plan (NRP), An Introduction

310-1 REQUIRED EXPERIENCE
Satisfactory performance as an Expanded Dispatch Supervisory Dispatcher (EDSP)
    +
Successful position performance as an Expanded Dispatch Coordinator (CORD)

310-1 PHYSICAL FITNESS LEVEL
None Required

310-1 OTHER POSITION ASSIGNMENTS THAT WILL MAINTAIN CURRENCY
Expanded Dispatch Supervisory Dispatcher (EDSP)

310-1 OTHER TRAINING WHICH SUPPORTS DEVELOPMENT OF KNOWLEDGE AND SKILLS
I-400 Incident Command System
L-480 Organizational Leadership in the Fire Service

Task Book available at:
http://www.nwcg.gov/pms/taskbook/dispatch/dispatch.htm

27.1 - Exhibit 01--Continued

POSITION QUALIFICATIONS SECTIONS
Dispatch Positions

EXPANDED DISPATCH SUPERVISORY DISPATCHER (EDSP)
Position Category: Incident Support

310-1 REQUIRED TRAINING
None

FOREST SERVICE ADDITIONAL REQUIRED TRAINING
D-510 Supervisory Dispatcher
IS-700 National Incident Management System (NIMS), An Introduction
IS-800 National Response Plan (NRP), An Introduction

310-1 REQUIRED EXPERIENCE
Satisfactory performance as an Expanded Dispatch Support Dispatcher (EDSD) in all four functional areas (Overhead, Crews, Equipment and Supplies)
    +
Successful position performance as an Expanded Dispatch Supervisory Dispatcher (EDSP)

310-1 PHYSCIAL FITNESS LEVEL
None Required

310-1 OTHER POSITION ASSIGNMENTS THAT WILL MAINTAIN CURRENCY
Expanded Dispatch Support Dispatcher (EDSD)
Any higher position for which this position is a prerequisite

310-1 OTHER TRAINING WHICH SUPPORTS DEVELOPMENT OF KNOWLEDGE AND SKILLS
A-207 Module, Aviation Conference and Education (ACE)
D-510 Supervisory Dispatcher
I-300 Intermediate Incident Command System
L-380 Fireline Leadership

Task Book available at:
http://www.nwcg.gov/pms/taskbook/dispatch/dispatch.htm

27.1 - Exhibit 01--Continued

POSITION QUALIFICATIONS SECTIONS
Dispatch Positions

EXPANDED DISPATCH SUPPORT DISPATCHER (EDSD)
Position Category: Incident Support

310-1 REQUIRED TRAINING
None

FOREST SERVICE ADDITIONAL REQUIRED TRAINING
D-310 Support Dispatcher
IS-700 National Incident Management System (NIMS), An Introduction
IS-800 National Response Plan (NRP), An Introduction

310-1 REQUIRED EXPERIENCE
Satisfactory performance as an Expanded Dispatch Recorder (EDRC)
        +
Successful position performance as an Expanded Dispatch Support Dispatcher (EDSD)

310-1 PHYSICAL FITNESS LEVEL
None

310-1 OTHER POSITION ASSIGNMENTS THAT WILL MAINTAIN CURRENCY
Expanded Dispatch Recorder (EDRC)
Supply Unit Leader (SPUL)
Any higher position for which this position is a prerequisite

310-1 OTHER TRAINING WHICH SUPPORTS DEVELOPMENT OF KNOWLEDGE AND SKILLS
D-310 Support Dispatcher
I-200 Basic Incident Command System
S-260 Interagency Incident Business Management
S-270 Basic Air Operations

Task Book available at:
http://www.nwcg.gov/pms/taskbook/dispatch/dispatch.htm

27.1 - Exhibit 01--Continued

POSITION QUALIFICATIONS SECTIONS
Dispatch Positions

EXPANDED DISPATCH RECORDER (EDRC)
Position Category: Incident Support

310-1 REQUIRED TRAINING
None

FOREST SERVICE ADDITIONAL REQUIRED TRAINING
I-100 Introduction to Incident Command System*
IS-700 National Incident Management System (NIMS), An Introduction

310-1 REQUIRED EXPERIENCE
Successful position performance as an Expanded Dispatch Recorder (EDRC)

310-1 PHYSICAL FITNESS LEVEL
None Required

310-1 OTHER POSITION ASSIGNMENTS THAT WILL MAINTAIN CURRENCY
Ordering Manager (ORDM)
Any higher position for which this position is a prerequisite

310-1 OTHER TRAINING WHICH SUPPORTS DEVELOPMENT OF KNOWLEDGE AND SKILLS
D-110 Dispatch Recorder
Basic Firefighter Training:
I-100 Introduction to Incident Command System
L-180 Human Factors on the Fireline
S-130 Firefighter Training
S-190 Introduction to Wildland Fire Behavior

Task Book available at:
http://www.nwcg.gov/pms/taskbook/dispatch/dispatch.htm
*I-100 Online link: *http://training.nwcg.gov/classes/i100.htm*

27.1 - Exhibit 01--Continued

POSITION QUALIFICATIONS SECTIONS
Dispatch Positions

INITIAL ATTACK DISPATCHER (IADP)
Position Category: Incident Support

310-1 REQUIRED TRAINING
Basic Firefighter Training:
I-100 Introduction to Incident Command System*
L-180 Human Factors on the Fireline
S-130 Firefighter Training
S-190 Introduction to Wildland Fire Behavior

FOREST SERVICE ADDITIONAL REQUIRED TRAINING
IS-700 National Incident Management System (NIMS), An Introduction
IS-800 National Response Plan (NRP), An Introduction

310-1 REQUIRED EXPERIENCE
Satisfactory performance as an Expanded Dispatch Recorder (EDRC)

    +

Successful position performance as an Initial Attack Dispatcher (IADP)

310-1 PHYSICAL FITNESS LEVEL
None Required

310-1 OTHER POSITION ASSIGNMENTS THAT WILL MAINTAIN CURRENCY
None

310-1 TRAINING WHICH SUPPORTS DEVELOPMENT OF KNOWLEDGE AND SKILLS
D-311 Initial Attack Dispatcher
S-200 Initial Attack Incident Commander
S-215 Fire Operations in Wildland/Urban Interface
S-271 Helicopter Crewmember
S-290 Intermediate Wildland Fire Behavior

Task Book available at:
http://www.nwcg.gov/pms/taskbook/dispatch/dispatch.htm
*I-100 Online link: *http://training.nwcg.gov/classes/i100.htm*

27.1 - Exhibit 01--Continued

POSITION QUALIFICATIONS SECTIONS
Dispatch Positions

AIRCRAFT DISPATCHER (ACDP)
Position Category: Incident Support

310-1 REQUIRED TRAINING
None

FOREST SERVICE ADDITIONAL REQUIRED TRAINING
D-312 Aircraft Dispatcher
IS-700 National Incident Management System (NIMS), An Introduction

310-1 REQUIRED EXPERIENCE
Satisfactory performance as an Expanded Dispatch Recorder (EDRC)
+
Successful position performance as an Aircraft Dispatcher (ACDP)

310-1 PHYSICAL FITNESS LEVEL
None Required

310-1 OTHER POSITION ASSIGNMENTS THAT WILL MAINTAIN CURRENCY
Expanded Dispatch Recorder (EDRC)
Aircraft Base Radio Operator (ABRO)

310-1 OTHER TRAINING WHICH SUPPORTS DEVELOPMENT OF KNOWLEDGE AND SKILLS
D-310 Support Dispatcher
I-200 Basic Incident Command System
S-260 Interagency Incident Business Management
S-270 Basic Air Operations

Task Book available at:
http://www.nwcg.gov/pms/taskbook/dispatch/dispatch.htm

27.1 - Exhibit 01--Continued

POSITION QUALIFICATIONS SECTIONS
Prevention & Investigation Positions

FIRE PREVENTION EDUCATION TEAM LEADER (PETL)
Position Category: Associated Activities

310-1 REQUIRED TRAINING
None

FOREST SERVICE ADDITIONAL REQUIRED TRAINING
IS-700 National Incident Management System (NIMS), An Introduction
IS-800 National Response Plan (NRP), An Introduction

310-1 REQUIRED EXPERIENCE
Satisfactory performance as a Fire Prevention Education team Member (PETM)
    +
Successful position performance as a Fire Prevention Education Team Leader (PETL)

310-1 PHYSICAL FITNESS LEVEL
None Required

310-1 OTHER POSITION ASSIGNMENTS THAT WILL MAINTAIN CURRENCY
Fire Prevention Team Member (PETM)

310-1 OTHER TRAINING WHICH SUPPORTS DEVELOPMENT OF KNOWLEDGE AND SKILLS
I-200 Basic Incident Command System
P-301 Wildland Fire Prevention Planning
P-410 Fire Prevention Education Team leader

Task Book available at:
http://www.nwcg.gov/pms/taskbook/education/education.htm

27.1 - Exhibit 01--Continued

POSITION QUALIFICATIONS SECTIONS
Prevention & Investigation Positions

FIRE PREVENTION EDUCATION TEAM MEMBER (PETM)
Position Category: Associated Activities

310-1 REQUIRED TRAINING
None

FOREST SERVICE ADDITIONAL REQUIRED TRAINING
I-100 Introduction to Incident Command System*
IS-700 National Incident Management System (NIMS), An Introduction

310-1 REQUIRED EXPERIENCE
Successful position performance as a Fire Prevention Education Team Member (PETM)

310-1 PHYSICAL FITNESS LEVEL
None Required

310-1 OTHER POSITION ASSIGNMENTS THAT WILL MAINTAIN CURRENCY
Any higher position for which this position is a prerequisite

310-1 OTHER TRAINING WHICH SUPPORTS DEVELOPMENT OF KNOWLEDGE AND SKILLS
FI-110 Wildland Fire Observations and Origin Scene Protection for First Responders
I-100 Introduction to Incident Command System
P-101 Introduction to Wildland Prevention
P-310 Fire Prevention Education Team Member
S-110 Basic Wildland Fire Orientation
S-130 Firefighter Training
S-190 Introduction to Wildland Fire Behavior

Task Book available at:
http://www.nwcg.gov/pms/taskbook/education/education.htm
*I-100 Online link: *http://training.nwcg.gov/classes/i100.htm*

27.1 - Exhibit 01--Continued

POSITION QUALIFICATIONS SECTIONS
Prevention & Investigation Positions

WILDLAND FIRE INVESTIGATION TEAM MEMBER (INTM)
Position Category: Associated Activities

310-1 REQUIRED TRAINING
FI-310 Wildland Fire Investigation Case Management

FOREST SERVICE ADDITIONAL REQUIRED TRAINING:
I-200 Basic Incident Command System
IS-700 National Incident Management System (NIMS), An Introduction
RT-130 Annual Fireline Safety Refresher

310-1 REQUIRED EXPERIENCE
Successful position performance as a Wildland Fire Investigator (INVF)
+
Satisfactory performance as a Wildland Fire Investigation Team Member (INTM)

FOREST SERVICE REQUIRED EXPERIENCE
Successful completion of a basic land management police training program or a criminal investigator training program conducted by the Federal Law Enforcement Training Center (FLETC)*

310-1 PHYSICAL FITNESS LEVEL
Light or completion of the 1.5 mile run/walk test portion of the Physical Efficiency Battery (PEB) at the 25th percentile or above.

310-1 OTHER POSITION ASSIGNMENTS THAT WILL MAINTAIN CURRENCY
None

27.1 - Exhibit 01--Continued

POSITION QUALIFICATIONS SECTIONS
Prevention & Investigation Positions

WILDLAND FIRE INVESTIGATION TEAM MEMBER (INTM) - CONTINUED
Position Category: Associated Activities

310-1 OTHER TRAINING WHICH SUPPORTS DEVELOPMENT OF KNOWLEDGE AND SKILLS
FI-311 Wildland Fire Investigation: Civil Case Managment
I-300 Intermediate Incident Command System
S-290 Intermediate Wildland Fire Behavior
Interviewing and Interrogation Training

Task Book available at:
http://www.nwcg.gov/pms/taskbook/education/education.htm

*Certification of a Wildland Fire Investigator who has met this required experience does not include completion of a position task book. Competency for the position task book can be granted by indicating the year of completion of the training programs provided.

27.1 - Exhibit 01--Continued

POSITION QUALIFICATIONS SECTIONS
Prevention & Investigation Positions

WILDLAND FIRE INVESTIGATOR (INVF)
Position Category: Associated Activities

310-1 REQUIRED TRAINING
FI-210 Wildfire Origin and Cause Determination

FOREST SERVICE ADDITIONAL REQUIRED TRAINING:
I-200 Basic Incident Command System
IS-700 National Incident Management System (NIMS), An Introduction
RT-130 Annual Fireline Safety Refresher

310-1 REQUIRED EXPERIENCE
Successful position performance as a Wildland Fire Investigator (INVF)

FOREST SERVICE REQUIRED EXPERIENCE
Forest Protection Officer Certification
  AND
Successful position performance as a Wildland Fire Investigator (INVF)
  OR
Successful completion of a basic land management police training program or a criminal investigator training program conducted by the Federal Law Enforcement Training Center (FLETC)*

310-1 PHYSICAL FITNESS LEVEL
Light or completion of the 1.5 mile run/walk test portion of the Physical Efficiency Battery (PEB) at the 25th percentile or above.

310-1 OTHER POSITION ASSIGNMENTS THAT WILL MAINTAIN CURRENCY
Any higher position for which this position is a prerequisite.

27.1 - Exhibit 01--Continued

POSITION QUALIFICATIONS SECTIONS
Prevention & Investigation Positions

WILDLAND FIRE INVESTIGATOR (INVF) - CONTINUED
Position Category: Associated Activities

310-1 OTHER TRAINING WHICH SUPPORTS DEVELOPMENT OF KNOWLEDGE AND SKILLS
FI-110 Wildland Fire Observation and Origin Scene Protection for First Responders
I-200 Basic Incident Command System
S-190 Introduction to Wildland Fire Behavior

Task Book available at:
http://www.nwcg.gov/pms/taskbook/education/education.htm

*Certification of a Wildland Fire Investigator who has met this required experience does not include completion of a position task book. Competency for the position can be granted by indicating the year of completion of the training programs provided.

FOREST SERVICE FIRE AND AVIATION
QUALIFICATIONS GUIDE

CHAPTER 2, PART 2- QUALIFICATIONS AND CERTIFICATION
TECHNICAL SPECIALISTS

Effective Date: February 28, 2011, Updated 6/10/2011

Table of Contents

.

2.7 - Technical Specialist Position Qualifications

Each agency is tasked by the NWCG Operations and Workforce Development Committee (OWDC) with establishing minimum certification and qualifications standards for technical specialist positions. Section 2.7 provides position qualifications including training requirements, experience, physical requirements, and other positions meeting currency requirements for the technical specialist position category.

NWCG does not provide minimum standards for these positions. These positions are not contained within the National Wildland Fire Qualification System Guide, 310-1. A listing of approved titles and position job codes of identified technical specialists can be found on the Incident Qualifications and Certification System website: *http://iqcs.nwcg.gov/*.

For positions that do not have a position task book, the Washington Office, Regional or Forest Qualification Review Committee shall review and recommend to the certifying official an individual's certification and qualification, based on objective factors such as performance evaluations and visual observation of performance of duties of the positions.

INDEX TO TECHNICAL SPECIALIST POSITIONS AND QUALIFICATIONS

INDEX TO TECHNICAL SPECIALIST POSITIONS AND QUALIFICATIONS

INDEX TO TECHNICAL SPECIALIST POSITIONS AND QUALIFICATIONS

ACCOUNTING TECHNICIAN (ACCT)

REQUIRED TRAINING
I-100 Introduction to Incident Command System*
I-200 Basic Incident Command
IS-700 National Incident Management System (NIMS), An Introduction
S-260 Interagency Incident Business Management

REQUIRED EXPERIENCE
On-the-job exposure to fire payment processing where applicable**
  OR
Voucher Examiner Knowledge or Experience
  OR
Unit Timekeeper

PHYSICAL FITNESS LEVEL
None

OTHER POSITION ASSIGNMENTS THAT WILL MAINTAIN CURRENCY
Buying Team Member (BUYM)

* Online training at: *http://training.nwcg.gov/classes/i100.htm*
** Requirements contained in Forest Service Handbook (FSH) 6509.13a, Assistant Disbursing Officer Handbook.

AERIAL OBSERVER (AOBS)

REQUIRED TRAINING
A-101 Aviation Safety (All Aircraft)
A-105 Aviation Life Support Equipment
A-106 Aviation Mishap Reporting
A-107 Aviation Policy and Regulations 1
A-109 Aviation Radio Use
A-113 Crash Survival
IS-700 National Incident Management System (NIMS), An Introduction

REQUIRED EXPERIENCE
For mobilization on the local unit, the Forest Qualification Review Committee shall determine the required experience necessary to perform in the position.

For mobilization off the unit:
Incident Commander Type 5 (ICT5)
     AND
Satisfactory performance as an Aerial Observer (AOBS)

PHYSICAL FITNESS LEVEL
None

OTHER POSITION ASSIGNMENTS THAT WILL MAINTAIN CURRENCY
None

OTHER TRAINING WHICH SUPPORTS DEVELOPMENT OF KNOWLEDGE AND SKILLS
I-200 Basic Incident Command System
S-270 Basic Air Operations
Local on-the-job orientation or Developed Aerial Observer Training

AGENCY AVIATION MILITARY LIAISON (AAML)

REQUIRED TRAINING
None

REQUIRED EXPERIENCE
Desirable experience is military background
    AND EITHER
Helicopter Operations Specialist
    OR
Helicopter Pilot Inspector*
    AND
Satisfactory performance as an Agency Aviation Military Liaison

PHYSICAL FITNESS LEVEL
None Required

OTHER POSITION ASSIGNMENTS THAT WILL MAINTAIN CURRENCY
Helicopter Operations Specialist
Helicopter Pilot Inspector (HPIN)

*References are contained in the Military Use Handbook (NFES 2175) located at the following website:
http://www.nifc.gov/nicc/predictive/intelligence/military/Military_Use_Handbook_2006_2.pdf

AIRTANKER BASE MANAGER (ATBM)

REQUIRED TRAINING
A-108 Preflight Checklist Briefing*
A-110 Aviation Transport of Hazardous Materials* (Triennial requirement)
A-112 Mission Planning and Flight Request Process*
A-115 Automated Flight Following*
A-116 General Awareness Security Training*
A-202 Interagency Aviation Organizations*
A-203 Basic Airspace*
A-204 Aircraft Capabilities and Limitations*
Aviation Business System Training*
I-200 Basic Incident Command System
S-260 Interagency Incident Business Management
Airtanker Base Manager Workshop (Triennial requirement)***

REQUIRED EXPERIENCE
Desirable skills include familiarity with the National Airtanker contract
    AND
Familiarity with the National Long Term Fire Retardant contract
    AND
Satisfactory performance as a Ramp Manager (RAMP)
    AND
Successful performance as an Airtanker Base Manager (ATBM)

PHYSICAL FITNESS LEVEL
None

OTHER POSITION ASSIGNMENTS THAT WILL MAINTAIN CURRENCY
Fixed Wing Base Manager (FWBM)
MAFFS Airtanker Base Manager (MABM)
OTHER TRAINING WHICH SUPPORTS DEVELOPMENT OF KNOWLEDGE AND SKILLS
Geographic Area Airtanker Base Manager Training
Geographic Area Mixmaster Training
Geographic Area Fixed Wing Base Manager Training
Contracting Officer Representative Training
*Online training at: www.iat.gov
**Online training at: http://www.fs.fed.us/business/abs/training.php
***Reference Interagency Airtanker Base Operations Guide at:
    https://fs.fed.us/fire/aviation/av_library/index.html

Taskbook for this position is located at: http://www.nwcg.gov/pms/taskbook-
    agency/index.htm

AIRCRAFT TIMEKEEPER (ATIM)

REQUIRED TRAINING
A-107 Aviation Policy and Regulations I*
Aviation Business System Training**
I-100 Introduction to Incident Command System***
IS-700 National Incident Management System (NIMS), An Introduction

REQUIRED EXPERIENCE
Satisfactory performance as an Aircraft Timekeeper (ATIM)

PHYSICAL FITNESS LEVEL
None Required

OTHER POSITION ASSIGNMENTS THAT WILL MAINTAIN CURRENCY
Airtanker Base Manager (ATBM)
Fixed Wing Base Manager (FWBM)
Helicopter Crewmember (HECM)
MAFFS Airtanker Base Manager (MABM)

OTHER TRAINING WHICH SUPPORTS DEVELOPMENT OF KNOWLEDGE AND SKILLS
A-104 Overview of Aircraft Capabilities and Limitations

*Online training at: http://www.iat.gov
**Online training at: *http://www.fs.fed.us/business/abs/training.php*
***Online training at: *http://training.nwcg.gov/classes/i100.htm*

ASSISTANT CACHE MANAGER (ACMR)

REQUIRED TRAINING
I-100 Introduction to Incident Command System*
IS-700 National Incident Management System (NIMS), An Introduction
S-260 Interagency Incident Business Management

REQUIRED CERTIFICATION
Hazmat Certification for 49 CFR

REQUIRED EXPERIENCE
Experience working in the National Cache System
    AND
Familiarity with the National Fire Equipment System (NFES)
    AND
Familiarity with the National Interagency Cache Business System (ICBS)
    AND
Satisfactory performance as Assistant Cache Manager (ACMR)

PHYSICAL FITNESS LEVEL
None

OTHER POSITION ASSIGNMENTS THAT WILL MAINTAIN CURRENCY
Cache Manager
Cache (Supply Clerk), Supervisory (CAST)
Material Handler Leader (WHLR)

OTHER TRAINING WHICH SUPPORTS DEVELOPMENT OF KNOWLEDGE AND SKILLS
I-200 Basic Incident Command System
I-300 Intermediate Incident Command

Task Book available at: *http://www.nwcg.gov/pms/taskbook/taskbook.htm*
* Online training at: *http://training.nwcg.gov/classes/i100.htm*

BATTALION MILITARY LIAISON (BNML)

REQUIRED TRAINING
RT-130 Annual Fireline Safety Refresher

REQUIRED EXPERIENCE
Desirable skills are ability to deal with individuals from multiple organizations and prior military experience
      AND
Operations Section Chief Type 1
      AND
Satisfactory performance as a Battalion Military Liaison (BNML)

PHYSICAL FITNESS LEVEL
Moderate

OTHER POSITION ASSIGNMENTS THAT WILL MAINTAIN CURRENCY
Operations Section Chief Type 1 (OSC1)

References are contained in the Military Use Handbook (NFES 2175)

BURNED AREA EMERGENCY RESPONSE SPECIALIST (BAES)

REQUIRED TRAINING
Burned Area Emergency Response Team Training
RT-130 Annual Fireline Safety Refresher
S-130 Firefighter Training
S-190 Introduction to Fire Behavior

REQUIRED EXPERIENCE
Satisfactory performance as a Burned Area Emergency Response Specialist (BAES)

PHYSICAL FITNESS LEVEL
Light

OTHER POSITION ASSIGNMENTS THAT WILL MAINTAIN CURRENCY
None

OTHER TRAINING WHICH SUPPORTS DEVELOPMENT OF KNOWLEDGE AND SKILLS
None

The Forest Service has not adopted the position standards established by the Department of Interior for other Burned Area Response Team members including:
BAER Team Leader (BAEL)
BAER Environmental Specialist (BAEN)
BAER Botanist (BABO)
BAER Documentation Specialist (BADO)
BAER Forester (BAFO)
BAER Soil Scientist (BASS)
BAER Hydrologist (BAHY)
BAER Geologist (BAEG)
BAER Biologist (BABI)
BAER Cultural Resource Specialist (BACS)

BUYING TEAM LEADER (BUYL)

REQUIRED TRAINING:
I-100 Introduction to Incident Command System*
IS-700 National Incident Management System (NIMS), An Introduction
S-260 Interagency Incident Business Management

REQUIRED CERTIFICATION
Recommended by Regional Director of Acquisitions

REQUIRED EXPERIENCE
Federal Delegated acquisition authority to obligate government funds of $100,000 or greater.

PHYSICAL FITNESS LEVEL
None

OTHER POSITION ASSIGNMENTS THAT WILL MAINTAIN CURRENCY
Procurement Unit Leader (PROC)

OTHER TRAINING WHICH SUPPORTS DEVELOPMENT OF KNOWLEDGE AND SKILLS
D-110 Dispatch Recorder
I-100 Introduction to Incident Command System
I-200 Basic Incident Command System
S-261 Applied Interagency Incident Business Management
S-360 Finance/Administrative Unit Leader
Regional Buying Team Workshop
National Interagency Buying Team Guide or Workshop
On-the-Job Procurement Training
Purchase Card and Convenience Check training National Interagency Buying Team Guide or Workshop

* Online training at: *http://training.nwcg.gov/classes/i100.htm*

BUYING TEAM MEMBER (BUYM)

REQUIRED TRAINING
I-100 Introduction to Incident Command System*
IS-700 National Incident Management System (NIMS), An Introduction
S-260 Interagency Incident Business Management

REQUIRED EXPERIENCE
Micro-purchase authority
  AND
Recommended by Regional Buying Team Coordinator

PHYSICAL FITNESS LEVEL
None

OTHER POSITION ASSIGNMENTS THAT WILL MAINTAIN CURRENCY
Buying Team Leader (BUYL)

OTHER TRAINING WHICH SUPPORTS DEVELOPMENT OF KNOWLEDGE AND SKILLS
National Interagency Buying Team Guide or Workshop
Regional Buying Team Workshop
On-the-Job Procurement Training
I-100 Introduction to Incident Command System
I-200 Basic Incident Command System
D-110 Dispatch Recorder
S-261 Applied Interagency Incident Business Management
Purchase Card and Convenience Check Training

* Online training at: *http://training.nwcg.gov/classes/i100.htm*

CACHE DEMOBILIZATION SPECIALIST (CDSP)

REQUIRED TRAINING
I-100 Introduction to Incident Command System*
IS-700 National Incident Management System (NIMS), An Introduction
L-180 Human Factors on the Fireline
National Cache Demobilization Specialist Training

REQUIRED CERTIFICATION
Hazmat Certification for 49 CFR

REQUIRED EXPERIENCE
Desirable skills include experience in receiving, accounting for and distributing supplies and familiarity with the National Fire Equipment System (NFES)
     AND
Satisfactory performance as a Cache Demobilization Specialist (CDSP)

PHYSICAL FITNESS LEVEL
None

OTHER POSITION ASSIGNMENTS THAT WILL MAINTAIN CURRENCY
Receiving/Distribution Manager (RCDM)
Assistant Cache Manager (ACMR)
Material Handler Leader (WHLR)

OTHER TRAINING WHICH SUPPORTS DEVELOPMENT OF KNOWLEDGE AND SKILLS
I-200 Basic Incident Command System
I-300 Intermediate Incident Command System

Task Book available at:
http://www.nwcg.gov/pms/taskbook/taskbook.htm

* Online training at: *http://training.nwcg.gov/classes/i100.htm*

CACHE SUPPLY CLERK (CASC)

REQUIRED TRAINING
I-100 Introduction to Incident Command System*
IS-700 National Incident Management System (NIMS), An Introduction
L-180 Human Factors on the Fireline

REQUIRED EXPERIENCE
Experience with computer applications and processing including the National Interagency
   Cache Business System (ICBS)
        AND
Familiarity with the National Fire Equipment System (NFES)
        AND
Satisfactory performance as a Cache Supply Clerk (CASC)

PHYSICAL FITNESS LEVEL
None

OTHER POSITION ASSIGNMENTS THAT WILL MAINTAIN CURRENCY
Cache (Supply) Clerk, Supervisory (CAST)

OTHER TRAINING WHICH SUPPORTS DEVELOPMENT OF KNOWLEDGE AND SKILLS
Cache (Supply) Clerk, Supervisory (CAST)

Task Book available at: *http://www.nwcg.gov/pms/taskbook/taskbook.htm*

* Online training at: *http://training.nwcg.gov/classes/i100.htm*

CACHE (SUPPLY) CLERK, SUPERVISORY (CAST)

REQUIRED TRAINING
I-100 Introduction to Incident Command System*
IS-700 National Incident Management System (NIMS), An Introduction

REQUIRED EXPERIENCE
Experience with computer applications and processing including the National Interagency Cache Business System (ICBS)
    AND
Familiarity with the National Fire Equipment System (NFES)
    AND
Cache (Supply) Clerk (CASC)
    AND
Satisfactory performance as Cache (Supply) Clerk, Supervisory (CAST)

PHYSICAL FITNESS
None

OTHER POSITION ASSIGNMENTS THAT WILL MAINTAIN CURRENCY
Cache (Supply) Clerk (CASC)

OTHER TRAINING WHICH SUPPORTS DEVELOPMENT OF KNOWLEDGE AND SKILLS
I-300 Intermediate Incident Command System

Task Book available at:
http://www.nwcg.gov/pms/taskbook/taskbook.htm

* Online training at: *http://training.nwcg.gov/classes/i100.htm*

CAMP CREW BOSS (CACB)

REQUIRED TRAINING
I-100 Introduction to Incident Command System*
IS-700 National Incident Management System (NIMS), An Introduction
S-260 Interagency Incident Business Management

REQUIRED EXPERIENCE
Satisfactory performance as a Camp Crew Boss

PHYSICAL FITNESS LEVEL
None

OTHER POSITION ASSIGNMENTS THAT WILL MAINTAIN CURRENCY
None

OTHER TRAINING WHICH SUPPORTS DEVELOPMENT OF KNOWLEDGE AND SKILLS
None

* Online training at: *http://training.nwcg.gov/classes/i100.htm*

COMMUNICATIONS COORDINATOR (COMC)

REQUIRED TRAINING
IS-700 National Incident Management System (NIMS), An Introduction
IS-800 National Response Framework (NRF), An Introduction
Communications Coordinator Course

REQUIRED EXPERIENCE
Communications Unit Leader (COML)
      AND
Satisfactory performance as a Communications Coordinator (COMC)

PHYSICAL FITNESS LEVEL
None

OTHER POSITION ASSIGNMENTS THAT WILL MAINTAIN CURRENCY
Communications Duty Officer at NIFC
Communications Unit Leader (COML) on a National Type 1 Team

OTHER TRAINING WHICH SUPPORTS DEVELOPMENT OF KNOWLEDGE AND SKILLS
None

Duties and responsibilities for this position can be found in chapter 60 of the National Interagency Mobilization Guide: *http://www.nifc.gov/news/mobguide/index.html*

CONTRACTING OFFICER (CONO)

REQUIRED TRAINING
I-100 Introduction to Incident Command System*
IS-700 National Incident Management System (NIMS), An Introduction
S-260 Interagency Incident Business Management
S-261 Applied Interagency Incident Business Management

REQUIRED AUTHORITY
Federal delegated acquisition authority to obligate Government funds of $100,000 or greater.

REQUIRED EXPERIENCE
None

PHYSICAL FITNESS
None

OTHER POSITION ASSIGNMENTS THAT WILL MAINTAIN CURRENCY
None

OTHER TRAINING WHICH SUPPORTS DEVELOPMENT OF KNOWLEDGE AND SKILLS
I-200 Basic Incident Command System
S-360 Finance/Administration Unit Leader
Incident Procurement Training

* Online training at: *http://training.nwcg.gov/classes/i100.htm*

COMPUTER COORDINATOR (COCO)

REQUIRED TRAINING
Computer Technical Specialist
I-100 Introduction to Incident Command System*
IS-700 National Incident Management System (NIMS), An Introduction

REQUIRED EXPERIENCE
Demonstrated skills include ability to set up, operate, and troubleshoot computer equipment problems
    AND
Satisfactory performance as a Computer Coordinator (COCO)

PHYSICAL FITNESS LEVEL
None

OTHER POSITION ASSIGNMENTS THAT WILL MAINTAIN CURRENCY
None

OTHER TRAINING WHICH SUPPORTS DEVELOPMENT OF KNOWLEDGE AND SKILLS
None

* Online training at: *http://training.nwcg.gov/classes/i100.htm*

COMPUTER DATA ENTRY RECORDER (CDER)

REQUIRED TRAINING
Incident Database Software Training
I-100 Introduction to Incident Command System*
IS-700 National Incident Management System (NIMS), An Introduction

REQUIRED EXPERIENCE
Demonstrated skills include proficiency in the use of word processing, database applications and communication software as well as experience working in a Windows environment.
          AND
Satisfactory performance as a Computer Data Entry Recorder (CDER)

PHYSICAL FITNESS LEVEL
None

OTHER POSITION ASSIGNMENTS THAT WILL MAINTAIN CURRENCY
None

OTHER TRAINING WHICH SUPPORTS DEVELOPMENT OF KNOWLEDGE AND SKILLS
None

* Online training at: *http://training.nwcg.gov/classes/i100.htm*

COMPUTER SPECIALIST (CTSP)

REQUIRED TRAINING
I-100 Introduction to Incident Command System*
IS-700 National Incident Management System (NIMS), An Introduction

REQUIRED EXPERIENCE
Demonstrated skills include ability to set up, operate, and troubleshoot computer equipment.
    AND
Satisfactory performance as a Computer Specialist (CTSP)

PHYSICAL FITNESS
None Required

OTHER POSITION ASSIGNMENTS THAT WILL MAINTAIN CURRENCY
Computer Coordinator

OTHER TRAINING WHICH SUPPORTS DEVELOPMENT OF KNOWLEDGE AND SKILLS
Computer Technical Specialist Training

* Online training at: *http://training.nwcg.gov/classes/i100.htm*

CONTRACTING OFFICER'S TECHNICAL REPRESENTATIVE (COTR)

REQUIRED TRAINING
I-100 Introduction to Incident Command System*
IS-700 National Incident Management System (NIMS), An Introduction
National Contracting Officer's Technical Representative Training

REQUIRED CERTIFICATION
Attendance at the National COTR workshop
    AND
Designated by National Contracting Officer

REQUIRED EXPERIENCE
Desirable skills are individuals who have completed Contracting Officer's Representative training.
    AND
Satisfactory performance as a Contracting Officer's Technical Representative

PHYSICAL FITNESS LEVEL
None

OTHER POSITION ASSIGNMENTS THAT WILL MAINTAIN CURRENCY
None

OTHER TRAINING WHICH SUPPORTS DEVELOPMENT OF KNOWLEDGE AND SKILLS
None

* Online training at: *http://training.nwcg.gov/classes/i100.htm*

DOZER OPERATOR (DZOP)

REQUIRED TRAINING
I-100 Introduction to Incident Command System*
IS-700 National Incident Management System (NIMS), An Introduction
RT-130 Annual Fireline Safety Refresher
S-130 Basic Firefighter
S-190 Basic Fire Behavior

CERTIFICATION
Dozer Operator Certification

REQUIRED EXPERIENCE
Satisfactory performance as a Dozer Operator (DZOP)

PHYSICAL FITNESS LEVEL
Moderate

OTHER POSITION ASSIGNMENTS THAT WILL MAINTAIN CURRENCY
Tractor Plow Operator

OTHER TRAINING WHICH SUPPORTS DEVELOPMENT OF KNOWLEDGE AND SKILLS
None

* Online training at: *http://training.nwcg.gov/classes/i100.htm*

DOZER OPERATOR INITIAL ATTACK (DZIA)

REQUIRED TRAINING
IS-700 National Incident Management System (NIMS), An Introduction
RT-130 Annual Fireline Safety Refresher
S-232 Dozer Boss
S-290 Intermediate Fire Behavior

CERTIFICATION
Dozer Operator Certification

REQUIRED EXPERIENCE
Dozer Operator (DZOP)
     AND
Satisfactory performance as a Dozer Operator Initial Attack (DZIA)

PHYSICAL FITNESS LEVEL
Moderate

OTHER POSITION ASSIGNMENTS THAT WILL MAINTAIN CURRENCY
Tractor/Plow Operator Initial Attack (TPIA)

OTHER TRAINING WHICH SUPPORTS DEVELOPMENT OF KNOWLEDGE AND SKILLS
None

EMERGENCY MEDICAL TECHNICIAN BASIC (EMTB)

REQUIRED TRAINING
IS-700 National Incident Management System (NIMS), An Introduction
S-130 Basic Firefighter
S-190 Introduction to Wildland Fire Behavior

CERTIFICATION
Current State (of origin) certification as an Emergency Medical Technician. May also require local certification in some jurisdictions.

REQUIRED EXPERIENCE
Knowledge and skills of the First Responder (uses a limited amount of equipment to perform initial assessment and intervention and is trained to assist other EMS providers) but is also qualified to function as minimum staff for an ambulance

PHYSICAL FITNESS LEVEL
None required

OTHER POSITION ASSIGNMENTS THAT WILL MAINTAIN CURRENCY
Advanced EMT (EMTA)
Paramedic. (EMTP)

OTHER TRAINING WHICH SUPPORTS DEVELOPMENT OF KNOWLEDGE AND SKILLS
Basic Trauma Life Support
Incident Medical Specialist Training
Cardiopulmonary Resuscitation (CPR) for the Professional Rescuer

EMERGENCY MEDICAL TECHNICIAN INTERMEDIATE (EMTI)

REQUIRED TRAINING
IS-700 National Incident Management System (NIMS), An Introduction

CERTIFICATION
Current State (of origin) certification as an EMTI. May also require local certification in some jurisdictions.

REQUIRED EXPERIENCE
Knowledge and skills of the preceding levels; in addition can perform essential advanced techniques and administer a limited number of medications

PHYSICAL FITNESS LEVEL
None Required

OTHER POSITION ASSIGNMENTS THAT WILL MAINTAIN CURRENCY
Paramedic (EMTP)

OTHER TRAINING WHICH SUPPORTS DEVELOPMENT OF KNOWLEDGE AND SKILLS
Basic Trauma Life Support (BTLS)
Advanced Trauma Life Support (ATLS)
Cardiopulmonary Resuscitation (CPR) for the Professional Rescuer

PARAMEDIC (EMTP)

REQUIRED TRAINING
IS-700 National Incident Management System (NIMS), An Introduction

REQUIRED LICENSE
Current State (of origin) certification as a Paramedic. May also require local certification in some jurisdictions.

REQUIRED EXPERIENCE
None

PHYSICAL FITNESS LEVEL
None required

OTHER POSITION ASSIGNMENTS THAT WILL MAINTAIN CURRENCY
None

OTHER TRAINING WHICH SUPPORTS DEVELOPMENT OF KNOWLEDGE AND SKILLS
Basic Trauma Life Support (BTLS)
Advanced Trauma Life Support (ATLS)
Cardiopulmonary Resuscitation (CPR) for the Professional Rescuer

ENGINE OPERATOR (ENOP)

REQUIRED TRAINING
IS-700 National Incident Management System (NIMS), An Introduction
RT-130 Annual Fireline Safety Refresher

REQUIRED LICENSE
Appropriate license and endorsements for make and model of engine
Training as required in FSM 5120.02 with regards to Lights & Siren use.

REQUIRED EXPERIENCE
Firefighter Type 1 (FFT1)
       AND
Satisfactory performance as an Engine Operator

PHYSICAL FITNESS
Arduous

OTHER POSITION ASSIGNMENTS THAT WILL MAINTAIN CURRENCY
Single Resource Boss Engine (ENGB)
Strike Team Leader Engine (STEN)

OTHER TRAINING WHICH SUPPORTS DEVELOPMENT OF KNOWLEDGE AND SKILLS
Geographic Area Engine Academy

Task Book available at:
http://www.nwcg.gov/pms/taskbook/taskbook.htm

FALLER CLASS A (FALA)

REQUIRED TRAINING
IS-700 National Incident Management System (NIMS), An Introduction
S-212 Wildland Fire Chain Saws
RT-212 Wildland Fire Chainsaw Refresher (Every three years after certification)

REQUIRED CERTIFICATION
 Initial chain saw certification and triennial re-certification

REQUIRED EXPERIENCE
Firefighter Type 2 (FFT2)

PHYSICAL FITNESS LEVEL
Arduous

OTHER POSITION ASSIGNMENTS THAT WILL MAINTAIN CURRENCY

OTHER TRAINING WHICH SUPPORTS DEVELOPMENT OF KNOWLEDGE AND SKILLS
None

The sawyer certification system outlined in FSH 6709.11, (sec. 22.48, b, 4(a-d)), and Regional Supplement, is used in the certification process for the Forest Service.

FALLER CLASS B (FALB)

REQUIRED TRAINING
IS-700 National Incident Management System (NIMS), An Introduction
S-212 Wildland Fire Chain Saws
RT-212 Wildland Fire Chainsaw Refresher (Every three years after initial training)

REQUIRED CERTIFICATION
Initial chain saw certification and triennial re-certification

REQUIRED EXPERIENCE:
Satisfactory performance as a Firefighter Type 2 (FFT2)

PHYSICAL FITNESS LEVEL
Arduous

OTHER POSITION ASSIGNMENTS THAT WILL MAINTAIN CURRENCY

OTHER TRAINING WHICH SUPPORTS DEVELOPMENT OF KNOWLEDGE AND SKILLS
None

The sawyer certification system outlined in FSH 6709.11, (sec. 22.48, b, 4(a-d)), and Regional Supplement, is used in the certification process for the Forest Service.

FALLER CLASS C (FALC)

REQUIRED TRAINING
IS-700 National Incident Management System (NIMS), An Introduction
Geographic Area Chainsaw Refresher (Initial and every three years)

REQUIRED CERTIFICATION
Initial chain saw certification and triennial re-certification

REQUIRED EXPERIENCE
Satisfactory performance as a Firefighter Type 2 (FFT2)

PHYSICAL FITNESS LEVEL
Arduous

OTHER POSITION ASSIGNMENTS THAT WILL MAINTAIN CURRENCY
None

OTHER TRAINING WHICH SUPPORTS DEVELOPMENT OF KNOWLEDGE AND SKILLS
None

The sawyer certification system outlined in FSH 6709.11, (sec. 22.48, b, 4(a-d)), Regional Supplement, is used in the certification process for the Forest Service.

FEMA EMERGENCY SUPPORT FUNCTION # 4 PRIMARY LEADER
(ESFL)

REQUIRED TRAINING
I-300 Intermediate Incident Command System
I-400 Advanced Incident Command System
IS-700 National Incident Management System (NIMS), An Introduction
IS-800 National Response Framework (NRF), An Introduction
ESF4 Training Course

REQUIRED CERTIFICATION
Regional Office FEMA Coordinator approval
Successful position performance as a Primary Leader ESFL

REQUIRED EXPERIENCE
US Forest Service Agency Personnel
    AND
Satisfactory position performance as a FEMA Emergency Support Function # 4 Primary
  Leader
    AND
Satisfactory performance as an Incident Command System Command or General Staff
  position at the Type 1 or 2 level
    OR
Geographic Area Coordinator
Regional/Area Fire Program Manager
Forest Fire Management Officer
Forest Supervisor
District Ranger

PHYSICAL FITNESS LEVEL
None Required

OTHER POSITION ASSIGNMENTS THAT WILL MAINTAIN CURRENCY
None

OTHER TRAINING WHICH SUPPORTS DEVELOPMENT OF KNOWLEDGE AND SKILLS
None

Task Book available at:
http://www.nwcg.gov/pms/taskbook-agency/index.htm

FEMA EMERGENCY SUPPORT FUNCTION # 4 WILDLAND SUPPORT
(ESFW)

REQUIRED TRAINING
I-300 Intermediate Incident Command System
I-400 Advanced Incident Command System
IS-700 National Incident Management System (NIMS), An Introduction
IS-800 National Response Framework (NRF), An Introduction
ESF4 Training Course

REQUIRED CERTIFICATION
Regional Office FEMA Coordinator approval
Successful position performance as a Wildland Support (EFSW)

REQUIRED EXPERIENCE
Satisfactory position performance as a FEMA Emergency Support Function # 4 Wildland
    Support (ESFW)
        AND
US Forest Service Forest Fire Management Officer,
US Forest Service Regional/Area Fire Program Specialist
Department of Interior Fire Management Officer
Supervisory Dispatcher

PHYSICAL FITNESS LEVEL
None Required

OTHER POSITION ASSIGNMENTS THAT WILL MAINTAIN CURRENCY
ESF4 Primary Leader (ESFL)

OTHER TRAINING WHICH SUPPORTS DEVELOPMENT OF KNOWLEDGE AND SKILLS
None

Taskbook available at:
http://www.nwcg.gov/pms/taskbook-agency/index.htm

FEMA EMERGENCY SUPPORT FUNCTION # 4 STRUCTURE SUPPORT (ESFS)

REQUIRED TRAINING
I-300 Intermediate Incident Command System
I-400 Advanced Incident Command System
IS-700 National Incident Management System (NIMS), An Introduction
IS-800 National Response Framework (NRF), An Introduction
ESF4 Training Course

REQUIRED EXPERIENCE
Served as Chief Officer in a Structure Fire Department
    OR
USFA Fire Program Specialist
USFA Training Specialist
    AND
Successful position performance as Structure Support (ESFS)

PHYSICAL FITNESS LEVEL
None Required

OTHER POSITION ASSIGNMENTS THAT WILL MAINTAIN CURRENCY
ESF4 Administrative Support (ESFA)

OTHER TRAINING WHICH SUPPORTS DEVELOPMENT OF KNOWLEDGE AND SKILLS
None

Taskbook available at:
http://www.nwcg.gov/pms/taskbook-agency/index.htm

FEMA EMERGENCY SUPPORT FUNCTION # 4 ADMINISTRATIVE SUPPORT (ESFA)

REQUIRED TRAINING
I-200 Basic Incident Command System
IS-700 National Incident Management System (NIMS), An Introduction
IS-800 National Response Framework (NRF), An Introduction
ESF4 Training Course

REQUIRED EXPERIENCE
Experience working with Incident Management Teams, interagency cooperators, and
    additional support organizations (local, county, State, Federal, National Guard, Military,
    Tribal Government, or FEMA)
        AND
Skill in word processing and spreadsheet Applications
        AND
Successful position performance as Administrative Support (ESFA)

PHYSICAL FITNESS LEVEL
None Required

OTHER POSITION ASSIGNMENTS THAT WILL MAINTAIN CURRENCY
None

OTHER TRAINING WHICH SUPPORTS DEVELOPMENT OF KNOWLEDGE AND SKILLS
None

Taskbook available at:
http://www.nwcg.gov/pms/taskbook-agency/index.htm

FIRE CACHE MANAGER (FCMG)

REQUIRED TRAINING
I-100 Introduction to Incident Command System*
IS-700 National Incident Management System (NIMS), An Introduction
32 Hours Supervisory training (USFS Corporate Training or L-380)

REQUIRED CERTIFICATION
Hazmat Certification for 49 CFR
      AND
Approved by Geographic Area Cache Manager

REQUIRED EXPERIENCE
Experience working within the National cache system
      AND
Experience with the National Interagency Cache Business System (ICBS)
      AND
Familiarity with the National Fire Equipment System (NFES)
      AND
Satisfactory performance as a Fire Cache Manager (FCMG)

PHYSICAL FITNESS
None Required

OTHER POSITION ASSIGNMENTS THAT WILL MAINTAIN CURRENCY
Assistant Fire Cache Manager (ACMR)

OTHER TRAINING WHICH SUPPORTS DEVELOPMENT OF KNOWLEDGE AND SKILLS
None

Task Book available at:
http://www.nwcg.gov/pms/taskbook/taskbook.htm
* Online training at: *http://training.nwcg.gov/classes/i100.htm*

FIRELINE EXPLOSIVES ADVISOR (FLEA)

REQUIRED TRAINING
IS-700 National Incident Management System (NIMS), An Introduction
RT-130 Annual Fireline Safety Refresher
Fireline Explosives and General Blaster Certification

REQUIRED CERTIFICATION
General Blaster qualified with Fireline Explosives Endorsement
> **AND**

Triennial re-certification
> **AND**

Commercial Drivers License

REQUIRED EXPERIENCE
Fireline Blaster (minimum of 2 complex and 3 total assignments as a Fireline Blaster)
> **AND**

Satisfactory performance as a Fireline Explosives Advisor (FLEA)

PHYSICAL FITNESS LEVEL
Light

OTHER POSITION ASSIGNMENTS THAT WILL MAINTAIN CURRENCY
None

OTHER TRAINING WHICH SUPPORTS DEVELOPMENT OF KNOWLEDGE AND SKILLS
None

FIRELINE EXPLOSIVES BLASTER (FLEB)

REQUIRED TRAINING
IS-700 National Incident Management System (NIMS), An Introduction
RT-130 Annual Fireline Safety Refresher
Fireline Explosives and General Blaster Certification

REQUIRED CERTIFICATION AND LICENSE
General Blaster qualified with Fireline Explosives endorsement
   AND
Triennial re-certification
   AND
Three active or dry firings per year
   AND
Commercial Drivers License

REQUIRED EXPERIENCE
Fireline Explosives Crewmember
   AND
Satisfactory performance as a Fireline Explosives Blaster (FLEB)

PHYSICAL FITNESS LEVEL
Arduous

OTHER POSITION ASSIGNMENTS THAT WILL MAINTAIN CURRENCY
Fireline Explosives Advisor (FLEA)

OTHER TRAINING WHICH SUPPORTS DEVELOPMENT OF KNOWLEDGE AND SKILLS
None

FIRELINE EXPLOSIVES CREWMEMBER (FLEC)

REQUIRED TRAINING
IS-700 National Incident Management System (NIMS), An Introduction
RT-130 Annual Fireline Safety Refresher
Fireline Explosives Training

REQUIRED CERTIFICATION
Triennial re-certification   **AND**
Three active or dry firings per year

REQUIRED EXPERIENCE
Satisfactory performance as a Fireline Explosives Crewmember (FLEC)

PHYSICAL FITNESS LEVEL
Arduous

OTHER POSITION ASSIGNMENTS THAT WILL MAINTAIN CURRENCY
Fireline Explosives Advisor (FLEA)
Fireline Explosives Blaster (FLEB)

OTHER TRAINING WHICH SUPPORTS DEVELOPMENT OF KNOWLEDGE AND SKILLS
None

FIXED WING BASE MANAGER (FWBM)

REQUIRED TRAINING
A-108 Preflight Checklist Briefing*
A-110 Aviation Transport of Hazardous Materials (Every three years)*
A-112 Mission Planning and Flight Request Process*
A-115 Automated Flight Following*
A-116 General Security Awareness*
A-200 Mishap Review*
A-202 Interagency Aviation Organizations*
A-203 Basic Airspace*
A-204 Aircraft Capabilities and Limitations*
Aviation Business System Training**
I-200 Basic Incident Command System
IS-700 National Incident Management System (NIMS), An Introduction
S-260 Interagency Incident Business Management

REQUIRED EXPERIENCE
Satisfactory performance as a Ramp Manager (RAMP)
    AND
Successful performance as a Fixed Wing Base Manager (FWBM)

PHYSICAL FITNESS LEVEL
None Required

OTHER POSITION ASSIGNMENTS THAT WILL MAINTAIN CURRENCY
Airtanker Base Manager (ATBM)
MAFFS Airtanker Base Manager (MABM)
Ramp Manager (RAMP)

OTHER TRAINING WHICH SUPPORTS DEVELOPMENT OF KNOWLEDGE AND SKILLS
Geographic Area Fixed Wing Base Manager training
Geographic Area Intermediate Air Operations

*Online training at: http://www.iat.gov
**Online training at: http://www.fs.fed.us/business/abs/training.php

Task book for this position is located at:
http://www.nwcg.gov/pms/taskbook-agency/index.htm

FIXED WING PARKING TENDER (FWPT)

REQUIRED TRAINING
A-101 Aviation Safety*
A-104 Overview of Aircraft Capabilities and Limitations*
A-109 Aviation Radio Use*
I-100 Introduction to Incident Command System**
IS-700 National Incident Management System (NIMS), An Introduction

REQUIRED EXPERIENCE
Satisfactory performance as a Fixed Wing Parking Tender (FWPT)

PHYSICAL FITNESS LEVEL
None Required

OTHER POSITION ASSIGNMENTS THAT WILL MAINTAIN CURRENCY
Airtanker Base Manager (ATBM)
Fixed Wing Base Manager (FWBM)
MAFFS Airtanker Base Manager (MABM)
Ramp Manager (RAMP)

OTHER TRAINING WHICH SUPPORTS DEVELOPMENT OF KNOWLEDGE AND SKILLS
Local Ramp Orientation

*Online training at: http://www.iat.gov
** Online training at: *http://training.nwcg.gov/classes/i100.htm*

FORWARD LOOKING INFRARED OPERATOR (FLIR)

REQUIRED TRAINING
I-100 Introduction to Incident Command System*
IS-700 National Incident Management System (NIMS), An Introduction
A-101 Aviation Safety (All Aircraft)
A-113 Crash Survival
A-204 Aircraft Capabilities & Limitations

REQUIRED EXPERIENCE
Satisfactory performance as a Forward Looking Infrared Operator (FLIR)

PHYSICAL FITNESS LEVEL
None Required

OTHER POSITION ASSIGNMENTS THAT WILL MAINTAIN CURRENCY
Infrared Interpreter (IRIN)

OTHER TRAINING WHICH SUPPORTS DEVELOPMENT OF KNOWLEDGE AND SKILLS
None

* Online training at: *http://training.nwcg.gov/classes/i100.htm*

HELICOPTER LONG LINE/REMOTE HOOK SPECIALIST (HELR)

REQUIRED TRAINING
A-110 Aviation Transport of Hazardous Materials (Must attend every three years)
I-100 Introduction to Incident Command System*
IS-700 National Incident Management System (NIMS), An Introduction
RT-130 Annual Fireline Safety Refresher
S-270 Basic Air Operations
> **AND**

A-219 Interagency Helicopter Transport of External Loads (Every three years)
> **OR**

S-271 Interagency Helicopter Crewmember

REQUIRED EXPERIENCE
Desirable skills are having worked around rotor wing aircraft (i.e., Helicopter Crewmember)
> **AND**

Satisfactory performance as a Helicopter Long Line/Remote Hook Specialist (HELR)

PHYSICAL FITNESS LEVEL
Arduous

OTHER POSITION ASSIGNMENTS THAT WILL MAINTAIN CURRENCY
Helicopter Crewmember (HECM)

OTHER TRAINING WHICH SUPPORTS DEVELOPMENT OF KNOWLEDGE AND SKILLS
None

* Online training at: *http://training.nwcg.gov/classes/i100.htm*

HELICOPTER RAPPEL SPOTTER (HERS)

REQUIRED TRAINING
IS-700 National Incident Management System (NIMS), An Introduction
RT-130 Annual Fireline Safety Refresher
FS-506 Helicopter Rappel Spotter
RT-506FS Annual Helicopter Rappel Spotter Refresher, required annually after initial training

REQUIRED EXPERIENCE
Satisfactory position performance as a Helicopter Manager (HMGB)
    AND
Satisfactory position performance as a Helicopter Rappel Spotter (HERS)

PHYSICAL FITNESS
Arduous

OTHER POSITION ASSIGNMENTS THAT WILL MAINTAIN CURRENCY
None

OTHER TRAINING WHICH SUPPORTS DEVELOPMENT OF KNOWLEDGE AND SKILLS
None

Reference materials are contained in the Interagency Helicopter Rappel Guide.

HELICOPTER RAPPELLER (HRAP)

REQUIRED TRAINING
I-100 Introduction to Incident Command System*
IS-700 National Incident Management System (NIMS), An Introduction
RT-130 Annual Fireline Safety Refresher
FS-505 Helicopter Rappeller
RT-505FS Annual Helicopter Rappeller Refresher

REQUIRED EXPERIENCE
Satisfactory performance as a Helicopter Rappeller (HRAP)
Satisfactory performance as a Helicopter Crewmember trainee (HECM t)

PHYSICAL FITNESS
Arduous

OTHER POSITION ASSIGNMENTS THAT WILL MAINTAIN CURRENCY
None

OTHER TRAINING WHICH SUPPORTS DEVELOPMENT OF KNOWLEDGE AND SKILLS
None

* Online training at: *http://training.nwcg.gov/classes/i100.htm*
Reference materials are contained in the Interagency Helicopter Rappel Guide.

HELITORCH MANAGER (HTMG)

REQUIRED TRAINING
A-110 Aviation Transport of Hazardous Material (Must attend every three years)
IS-700 National Incident Management System (NIMS), An Introduction
RT-130 Annual Fireline Safety Refresher
N-9012 Helitorch Manager
RT-9012 Annual Helitorch Manager Refresher, required annually after initial training.
Training in the appropriate ignition devices

REQUIRED EXPERIENCE
Satisfactory performance as a Helitorch Mixmaster (HTMM)
    AND
Satisfactory performance as a Helitorch Manager (HTMG)

PHYSICAL FITNESS LEVEL
Moderate

OTHER POSITION ASSIGNMENTS THAT WILL MAINTAIN CURRENCY
Helitorch Mixmaster (HTMM)

OTHER TRAINING WHICH SUPPORTS DEVELOPMENT OF KNOWLEDGE AND SKILLS
None

Reference materials are contained in the Interagency Aerial Ignition Guide (NFES 1080).

HELITORCH MIXMASTER (HTMM)

REQUIRED TRAINING:
A-110 Aviation Transport of Hazardous Material (Must attend every three years)
I-100 Introduction to Incident Command System*
IS-700 National Incident Management System (NIMS), An Introduction
RT-130 Annual Fireline Safety Refresher
N-9012 Helitorch Manager
RT-9012 Annual Helitorch Refresher, required annually after initial training

REQUIRED EXPERIENCE
Satisfactory performance as a Helitorch Mixmaster (HTMM)

PHYSICAL FITNESS
Light

OTHER POSITION ASSIGNMENTS THAT WILL MAINTAIN CURRENCY
None

OTHER TRAINING WHICH SUPPORTS DEVELOPMENT OF KNOWLEDGE AND SKILLS
None

* Online training at: *http://training.nwcg.gov/classes/i100.htm*
Reference materials are contained in the Interagency Aerial Ignition Guide (NFES 1080).

HELITORCH PARKING TENDER (HTPT)

REQUIRED TRAINING:
A-110 Aviation Transport of Hazardous Material (Every three years)
IS-700 National Incident Management System (NIMS), An Introduction
RT-130 Annual Fireline Safety Refresher
S-271 Interagency Helicopter Crewmember
N-9012 Helitorch Manager
RT-9012 Annual Helitorch Refresher, required annually after initial training

REQUIRED EXPERIENCE
Satisfactory position performance as a Helicopter Crewmember (HECM)
          AND
Satisfactory performance as a Helitorch Parking Tender (HTPT)

PHYSICAL FITNESS LEVEL
Moderate

OTHER POSITION ASSIGNMENTS THAT WILL MAINTAIN CURRENCY
None

OTHER TRAINING WHICH SUPPORTS DEVELOPMENT OF KNOWLEDGE AND SKILLS
None

Reference materials are contained in the Interagency Aerial Ignition Guide (NFES 1080).

INCIDENT MEDICAL SPECIALIST ASSISTANT (IMSA)

REQUIRED TRAINING
I-100 Introduction to Incident Command System*
IS-700 National Incident Management System (NIMS), An Introduction
N9013 Geographic Area Medical Specialist Workshop

REQUIRED CERTIFICATION
Current State EMT Certification
Documented Hepatitis B Vaccination
Attendance at Geographic Area Medical Specialist Workshop/Training

REQUIRED EXPERIENCE
Satisfactory performance as a Incident Medical Technician (IMST)
 AND
Active with an Emergency Medical Provider
 AND
Satisfactory performance as an Incident Medical Assistant (IMSA)

PHYSICAL FITNESS LEVEL
Light

OTHER POSITION ASSIGNMENTS THAT WILL MAINTAIN CURRENCY
None

OTHER TRAINING WHICH SUPPORTS DEVELOPMENT OF KNOWLEDGE AND SKILLS
None

* Online training at: *http://training.nwcg.gov/classes/i100.htm*

INCIDENT MEDICAL MANAGER (IMSM)

REQUIRED TRAINING
I-100 Introduction to Incident Command System*
IS-700 National Incident Management System (NIMS), An Introduction
N9013 Geographic Area Medical Specialist Workshop

REQUIRED CERTIFICATION
Current State EMT Certification
Documented Hepatitis B Vaccination
Attendance at Geographic Area Medical Specialist Workshop/Training

REQUIRED EXPERIENCE
Satisfactory performance as an Incident Medical Assistant (IMSA)
      AND
Active with an Emergency Medical Provider
      AND
Satisfactory performance as an Incident Medical Manager (IMSM)

PHYSICAL FITNESS LEVEL
Light

OTHER POSITION ASSIGNMENTS THAT WILL MAINTAIN CURRENCY
None

OTHER TRAINING WHICH SUPPORTS DEVELOPMENT OF KNOWLEDGE AND SKILLS
None

* Online training at: *http://training.nwcg.gov/classes/i100.htm*

INCIDENT MEDICAL TECHNICIAN (IMST)

REQUIRED TRAINING
I-100 Introduction to Incident Command System*
I-200 Basic Incident Command System
IS-700 National Incident Management System (NIMS), An Introduction

REQUIRED CERTIFICATION
Current State EMT Certification
Documented Hepatitis B Vaccination
Biennial attendance at RT-9013 Geographic Area Incident Medical Specialist Training

REQUIRED EXPERIENCE
Satisfactory performance as an Incident Medical Technician (IMST)

PHYSICAL FITNESS LEVEL
Light

OTHER POSITION ASSIGNMENTS THAT WILL MAINTAIN CURRENCY
None

OTHER TRAINING WHICH SUPPORTS DEVELOPMENT OF KNOWLEDGE AND SKILLS
None

* Online training at: *http://training.nwcg.gov/classes/i100.htm*

INCIDENT METEOROLOGIST (IMET)

REQUIRED TRAINING
I-100 Introduction to Incident Command System*
IS-700 National Incident Management System (NIMS), An Introduction
S-130 Basic Firefighter
S-190 Introduction to Fire Behavior
S-290 Intermediate Fire Behavior

REQUIRED EXPERIENCE
Meteorologist
     AND
Satisfactory performance as an Incident Meteorologist (IMET)

PHYSICAL FITNESS LEVEL
None Required

OTHER POSITION ASSIGNMENTS THAT WILL MAINTAIN CURRENCY
None

OTHER TRAINING WHICH SUPPORTS DEVELOPMENT OF KNOWLEDGE AND SKILLS
S-390 Introduction to Wildland Fire Behavior Calculations
S-490 Advanced Wildland Fire Behavior Calculations

* Online training at: *http://training.nwcg.gov/classes/i100.htm*

INFRARED DOWNLINK OPERATOR (IRDL)

REQUIRED TRAINING
IS-700 National Incident Management System (NIMS), An Introduction

REQUIRED EXPERIENCE
Satisfactory performance as an Infrared Interpreter (IRIN)
    AND
Satisfactory performance as an Infrared Downlink Operator (IRDL)

PHYSICAL FITNESS LEVEL
None Required

OTHER POSITION ASSIGNMENTS THAT WILL MAINTAIN CURRENCY
None

OTHER TRAINING WHICH SUPPORTS DEVELOPMENT OF KNOWLEDGE AND SKILLS
None

INFRARED FIELD SPECIALIST (IRFS)

REQUIRED TRAINING
I-100 Introduction to Incident Command System*
IS-700 National Incident Management System (NIMS), An Introduction
S-443 Infrared Interpreter

REQUIRED EXPERIENCE
Satisfactory performance as an Infrared Field Specialist (IRFS)

PHYSICAL FITNESS LEVEL
Light

OTHER POSITION ASSIGNMENTS THAT WILL MAINTAIN CURRENCY
None

OTHER TRAINING WHICH SUPPORTS DEVELOPMENT OF KNOWLEDGE AND SKILLS
None

* Online training at: *http://training.nwcg.gov/classes/i100.htm*

INFRARED INTERPRETER (IRIN)

REQUIRED TRAINING
I-100 Introduction to Incident Command System*
IS-700 National Incident Management System (NIMS), An Introduction
S-443 Infrared Interpreter

REQUIRED EXPERIENCE
Satisfactory performance as an Infrared Interpreter (IRIN)

PHYSICAL FITNESS
None Required

OTHER POSITION ASSIGNMENTS THAT WILL MAINTAIN CURRENCY
None

OTHER TRAINING WHICH SUPPORTS DEVELOPMENT OF KNOWLEDGE AND SKILLS
None

Taskbook available at: *http://www.nwcg.gov/pms/taskbook/taskbook.htm*
* Online training at: *http://training.nwcg.gov/classes/i100.htm*

INFRARED REGIONAL COORDINATOR (IRCR)

REQUIRED TRAINING
IS-700 National Incident Management System (NIMS), An Introduction

REQUIRED CERTIFICATION
Designated, annually, by the National Infrared Interpreter Program Manager

REQUIRED EXPERIENCE
Satisfactory performance as an Infrared Interpreter (IRIN)
    AND
Satisfactory performance as an Infrared Regional Coordinator (IRCR)

PHYSICAL FITNESS LEVEL
None Required

OTHER POSITION ASSIGNMENTS THAT WILL MAINTAIN CURRENCY
Infrared Interpreter (IRIN)

OTHER TRAINING WHICH SUPPORTS DEVELOPMENT OF KNOWLEDGE AND SKILLS
None

INTELLIGENCE SUPPORT (INTS)

REQUIRED TRAINING
I-100 Introduction to Incident Command System*
IS-700 National Incident Management System (NIMS), An Introduction

REQUIRED EXPERIENCE
Satisfactory performance as an Intelligence Support

PHYSICAL FITNESS LEVEL
None Required

OTHER POSITION ASSIGNMENTS THAT WILL MAINTAIN CURRENCY
None

OTHER TRAINING WHICH SUPPORTS DEVELOPMENT OF KNOWLEDGE AND SKILLS
Intelligence Support (INTS) Training
Weather Information Management System (WIMS)
Entry Level National Fire Danger Rating System (NFDRS)
Resource Order Status System (ROSS) Computer Based Training (Reports)
Spreadsheet (Excel) and Database (Access) Software
D-110 Dispatch Recorder
I-200 Basic Incident Command System
S-130 Firefighter Training
S-190 Introduction to Wildland Fire Behavior

* Online training at: *http://training.nwcg.gov/classes/i100.htm*
Taskbook available at: *http://www.nwcg.gov/pms/taskbook/taskbook.htm*

MAC GROUP COORDINATOR (MCCO)

REQUIRED TRAINING:
IS-700 National Incident Management System (NIMS), An Introduction
IS-800 National Response Framework (NRF), An Introduction

REQUIRED EXPERIENCE
Satisfactory performance as a MAC Group Coordinator

PHYSICAL FITNESS LEVEL
None Required

OTHER POSITION ASSIGNMENTS THAT WILL MAINTAIN CURRENCY
None

OTHER TRAINING WHICH SUPPORTS DEVELOPMENT OF KNOWLEDGE AND SKILLS
I-400 Advanced Incident Command System
M-480 Multi-Agency Coordinating Group

MAC GROUP INFORMATION OFFICER (MCIF)

REQUIRED TRAINING
IS-700 National Incident Management System (NIMS), An Introduction
IS-800 National Response Framework (NRF), An Introduction
I-300 Intermediate Incident Command System

REQUIRED EXPERIENCE
Satisfactory performance as a Public Information Officer. Type 2 (PIO2)
    AND
Satisfactory performance as a MAC Group Information Officer (MCIF)

PHYSICAL FITNESS LEVEL
None Required

OTHER POSITION ASSIGNMENTS THAT WILL MAINTAIN CURRENCY
Public Information Officer, Type 1 (PIO1)

OTHER TRAINING WHICH SUPPORTS DEVELOPMENT OF KNOWLEDGE AND SKILLS
I-400 Advanced Incident Command System
M-480 Multi-Agency Coordinating Group

MAFFS AIRTANKER BASE MANAGER (MABM)

REQUIRED TRAINING
MAFFS Training Exercise or MAFFS Activation (recurrent training)

REQUIRED EXPERIENCE
Satisfactory performance as an Airtanker Base Manager (ATBM)
    AND
Satisfactory performance as a MAFFS Tanker Base Specialist (MABS)
    AND
Satisfactory performance as a MABM Trainee at MAFFS Training (Ref. MAFFS Ops. Plan)
    AND
Satisfactory performance as a MAFFS Airtanker Base Manager (MABM)

PHYSICAL FITNESS LEVEL
None

OTHER POSITION ASSIGNMENTS THAT WILL MAINTAIN CURRENCY
MAFFS Tanker Base Specialist (MABS)

OTHER TRAINING WHICH SUPPORTS DEVELOPMENT OF KNOWLEDGE AND SKILLS
None

Reference materials for this position are contained within the MAFFS Operating Plan.

MAFFS AIRTANKER BASE SPECIALIST (MABS)

REQUIRED TRAINING
IS-700 National Incident Management System (NIMS), An Introduction
MAFFS Training Exercise

REQUIRED EXPERIENCE
Satisfactory performance in any Airbase position
    OR
Satisfactory performance as a Single Engine Airtanker Manager (SEAT)
    AND
Satisfactory performance as a MAFFS Airtanker Base Specialist (MABS)

PHYSICAL FITNESS LEVEL
None

OTHER POSITION ASSIGNMENTS THAT WILL MAINTAIN CURRENCY
MAFFS Airtanker Base Manager

OTHER TRAINING WHICH SUPPORTS DEVELOPMENT OF KNOWLEDGE AND SKILLS
None

Reference materials for this position are contained within the MAFFS Operating Plan.

MAFFS CLERK (MAFC)

REQUIRED TRAINING
I-100 Introduction to Incident Command System*
IS-700 National Incident Management System (NIMS), An Introduction

REQUIRED EXPERIENCE
Desirable skills are record keeping, organization ability and communication skills
　　AND
Satisfactory performance as a MAFFS Clerk (MAFC)

PHYSICAL FITNESS LEVEL
None Required

OTHER POSITION ASSIGNMENTS THAT WILL MAINTAIN CURRENCY
Expanded Dispatch Recorder (EDRC)
Documentation Unit Leader (DOCL)

OTHER TRAINING WHICH SUPPORTS DEVELOPMENT OF KNOWLEDGE AND SKILLS
None

* Online training at: *http://training.nwcg.gov/classes/i100.htm*

MAFFS LIAISON OFFICER (MAFF)

REQUIRED TRAINING
MAFFS Liaison Officer Training

CERTIFICATION
Approved by the National Military Liaison Officer

REQUIRED EXPERIENCE
Prior military experience is desirable, but not a required prerequisite
    AND
Satisfactory performance as a MAFFS Liaison Officer (MAFF)

PHYSICAL FITNESS LEVEL
None Required

OTHER POSITION ASSIGNMENTS THAT WILL MAINTAIN CURRENCY
None

OTHER TRAINING WHICH SUPPORTS DEVELOPMENT OF KNOWLEDGE AND SKILLS
None

Reference materials for this position are in the MAFFS Operating Plan.

MILITARY AVIATION OPERATIONS COORDINATOR (MAOC)

REQUIRED TRAINING
None

REQUIRED CERTIFICATION
Approved by National Helicopter Specialist

REQUIRED EXPERIENCE
Qualified as a Helicopter Operations Specialist
> **OR**

Qualified as a Helicopter Pilot Inspector (HPIN)
> **AND**

Satisfactory performance as a Military Aviation Operations Coordinator (MAOC)

PHYSICAL FITNESS LEVEL
None Required

OTHER POSITION ASSIGNMENTS THAT WILL MAINTAIN CURRENCY
None

OTHER TRAINING WHICH SUPPORTS DEVELOPMENT OF KNOWLEDGE AND SKILLS
None

Reference materials for this position are in the Military Use Handbook (NFES 2175).

MILITARY CREW LIAISON ADVISOR (MCAD)

REQUIRED TRAINING
IS-700 National Incident Management System (NIMS), An Introduction
RT-130 Annual Fireline Safety Refresher

REQUIRED EXPERIENCE
Prior military experience is desirable, but not a required prerequisite
    AND
Successful position performance as a Single Resource Boss Crew (CRWB)
    AND
Satisfactory performance as a Military Crew Liaison Advisor (MCAD)

PHYSICAL FITNESS LEVEL
Arduous

OTHER POSITION ASSIGNMENTS THAT WILL MAINTAIN CURRENCY
Single Resource Boss Crew (CRWB)

OTHER TRAINING WHICH SUPPORTS DEVELOPMENT OF KNOWLEDGE AND SKILLS
None

Reference materials for this position are in the Military Use Handbook (NFES 2175).

MIXMASTER (MXMS)

REQUIRED TRAINING
S-270 Basic Air Operations

REQUIRED EXPERIENCE
Satisfactory performance as a Retardant Crewmember
  AND
Satisfactory performance as a Mixmaster (MXMS)

PHYSICAL FITNESS
None Required

OTHER POSITION ASSIGNMENTS THAT WILL MAINTAIN CURRENCY
Airtanker Base Manager (ATBM)
MAFFS Airtanker Base Manager MABM

OTHER TRAINING WHICH SUPPORTS DEVELOPMENT OF KNOWLEDGE AND SKILLS
Geographic Area Mixmaster Training

PLASTIC SPHERE DISPENSER OPERATOR (PLDO)

REQUIRED TRAINING
A-110 Aviation Transport of Hazardous Material (Must attend every three years)
IS-700 National Incident Management System (NIMS), An Introduction
RT-130 Annual Fireline Safety Refresher
N-9016 Plastic Sphere Dispenser
RT-9016 Annual Plastic Sphere Dispenser Refresher, required annually after initial training

REQUIRED EXPERIENCE
Satisfactory performance as a Helicopter Crewmember (HECM)

PHYSICAL FITNESS LEVEL
None Required

OTHER POSITION ASSIGNMENTS THAT WILL MAINTAIN CURRENCY
None

OTHER TRAINING WHICH SUPPORTS DEVELOPMENT OF KNOWLEDGE AND SKILLS
Geographic Area Mixmaster Training

Reference materials are contained in the Interagency Aerial Ignition Guide (NFES 1080).

PURCHASING AGENT, FIVE THOUSAND (PA05)
PURCHASING AGENT, TEN THOUSAND (PA10)
PURCHASING AGENT, TWENTY-FIVE THOUSAND (PA25)
PURCHASING AGENT, FIFTY THOUSAND (PA50)

REQUIRED TRAINING
I-100 Introduction to Incident Command System*
IS-700 National Incident Management System (NIMS), An Introduction

REQUIRED EXPERIENCE
Federal delegated acquisition authority to obligate Government funds and appropriate authority for the positions.

PHYSICAL FITNESS LEVEL
None Required

OTHER POSITION ASSIGNMENTS THAT WILL MAINTAIN CURRENCY
None

OTHER TRAINING WHICH SUPPORTS DEVELOPMENT OF KNOWLEDGE AND SKILLS
I-200 Basic Incident Command System
S-260 Interagency Incident Business Management
S-261 Applied Interagency Incident Business Management
S-360 Finance/Administration Unit Leader

* Online training at: *http://training.nwcg.gov/classes/i100.htm*

PRESCRIBED FIRE BURN BOSS TYPE 3 (RXB3)

REQUIRED TRAINING:
IS-700 National Incident Management System (NIMS), An Introduction
RT-130 Annual Fireline Safety Refresher
S-290 Intermediate Wildland Fire Behavior

REQUIRED EXPERIENCE
Satisfactory performance as a Firefighter Type 1 (FFT1)
    AND
Satisfactory performance as a Prescribed Fire Burn Boss Type 3 (RXB3)

PHYSICAL FITNESS
Moderate

OTHER POSITION ASSIGNMENTS THAT WILL MAINTAIN CURRENCY
Prescribed Fire Burn Boss Type 2 (RXB2)
Prescribed Fire Burn Boss Type 1 (RXB1)
Strategic Operational Planner (SOPL)

OTHER TRAINING WHICH SUPPORTS DEVELOPMENT OF KNOWLEDGE AND SKILLS
S-234 Ignition Operations

Task Book is available at: *http://www.fs.fed.us/fire/fireuse/rxfire/rxb3_ptb.pdf*

PRESCRIBED FIRE CREWMEMBER (RXCM)

REQUIRED TRAINING
I-100 Introduction to Incident Command System*
IS-700 National Incident Management System (NIMS), An Introduction
RT-130 Annual Fireline Safety Refresher
S-130 Firefighter Training
S-190 Introduction to Wildland Fire Behavior

REQUIRED EXPERIENCE
None

PHYSICAL FITNESS LEVEL
Moderate

OTHER POSITION ASSIGNMENTS THAT WILL MAINTAIN CURRENCY
Firefighter Type 2 (FFT2)

OTHER TRAINING WHICH SUPPORTS DEVELOPMENT OF KNOWLEDGE AND SKILLS
S-211 Portable Pumps and Water Use
S-234 Ignition Operations

The RXCM serves as a member of a crew working under the immediate supervision of a qualified burn boss (as described in this section) on low/moderate complexity prescribed burns in moderate terrain.

* Online training at: *http://training.nwcg.gov/classes/i100.htm*

RAMP MANAGER (RAMP)

REQUIRED TRAINING
A-105 Aviation Life Support Equipment*
A-106 Aviation Mishap Reporting*
A-107 Aviation Policy and Regulations I*
S-270 Basic Air Operations

REQUIRED EXPERIENCE
Satisfactory performance as a Fixed Wing Parking Tender (FWPT)
       AND
Satisfactory performance as a Ramp Manager (RAMP)

PHYSICAL FITNESS LEVEL
None Required

OTHER POSITION ASSIGNMENTS THAT WILL MAINTAIN CURRENCY
Airtanker Base Manager (ATBM)
Fixed Wing Base Manager (FWBM)
Fixed Wing Parking Tender (FWPT)
MAFFS Airtanker Base Manager (MABM)

OTHER TRAINING WHICH SUPPORTS DEVELOPMENT OF KNOWLEDGE AND SKILLS
A-110Aviation Transport of Hazardous Materials*
A-204 Aircraft Capabilities and Limitations*

Online training located at: http://www.iat.gov

REMOTE AUTOMATED WEATHER STATION TECHNICIAN (RAWS)

REQUIRED TRAINING:
National Approved RAWS Maintenance Training
I-100 Introduction to Incident Command System*
I-200 Basic Incident Command System
IS-700 National Incident Management System (NIMS), An Introduction
RT-130 Annual Fireline Safety Refresher
S-130 Basic Firefighter
S-190 Introduction to Wildland Fire Behavior

REQUIRED EXPERIENCE
Proficiency with the ASCADS System and experience with RAWS and REMS equipment
AND
Knowledge of fire weather observation procedures and weather station location
recommendations (per PMS 426-1 and 426-3 and the Fire Weather Observers' Handbook
#494, 1976)

PHYSICAL FITNESS LEVEL
Light

OTHER POSITION ASSIGNMENTS THAT WILL MAINTAIN CURRENCY
None

OTHER TRAINING WHICH SUPPORTS DEVELOPMENT OF KNOWLEDGE AND SKILLS
S-258 Communications Technician

* Online training at: *http://training.nwcg.gov/classes/i100.htm*

RETARDANT CREWMEMBER
(RTCM)

REQUIRED TRAINING
A-101 Aviation Safety*
A-104 Overview of Aircraft Capabilities and Limitations*
I-100 Introduction to Incident Command System**
IS-700 National Incident Management System (NIMS), An Introduction

REQUIRED EXPERIENCE
Satisfactory performance as a Retardant Crewmember

PHYSICAL FITNESS LEVEL
None Required

OTHER POSITION ASSIGNMENTS THAT WILL MAINTAIN CURRENCY
Mixmaster (MXMS)

OTHER TRAINING WHICH SUPPORTS DEVELOPMENT OF KNOWLEDGE AND SKILLS
Geographic Area Mixmaster Training

*Online training at: *http://www.iat.gov*
** Online training at: *http://training.nwcg.gov/classes/i100.htm*

SECURITY GUARD, NOT LAW ENFORCEMENT (SECG)

POSITION DESCRIPTION
Personnel utilized in this position shall not exercise law enforcement duties of either state or federal law, including arrest or detention of persons, nor carry weapons or other defensive equipment.

Uniforms may be worn and marked vehicles driven, however they shall not contain the words "police" or equivalent, or contain markings of a public law enforcement or police agency.

REQUIRED TRAINING
I-100 Introduction to Incident Command System*
IS-700 National Incident Management System (NIMS), An Introduction
Training as required within the state of the incident for peace officer status or security guard licensing as appropriate.

CERTIFICATION
Licensed and in compliance with any applicable requirements for security guards within the state of the respective incident.

Where state law allows, peace officer or law enforcement officer training or employment may meet requirements.

REQUIRED EXPERIENCE
Satisfactory performance as a Security Guard (SECG)

PHYSICAL FITNESS LEVEL
None Required

OTHER POSITION ASSIGNMENTS THAT WILL MAINTAIN CURRENCY
None

AD hiring authority and procurement of private services may be used for this position.
* Online training at: *http://training.nwcg.gov/classes/i100.htm*

SECURITY SPECIALIST LEVEL 1 (SEC1)

REQUIRED TRAINING:
Federal Law Enforcement Training Center (FLETC) Criminal Investigator or Land
Management Police Training Programs
I-100 Introduction to Incident Command System*
IS-700 National Incident Management System (NIMS), An Introduction

CERTIFICATION
Certification as Criminal Investigator or Law Enforcement Officer

AUTHORITY
Authorized and equipped to carry firearms, make arrests, serve warrants, conduct searches
and seizures. Authorized to enforce federal or state laws.

REQUIRED EXPERIENCE
Satisfactory position performance as a Security Specialist Level 1 (SEC1)

PHYSICAL FITNESS LEVEL
None Required

OTHER POSITION ASSIGNMENTS THAT WILL MAINTAIN CURRENCY
Security Manager (SECM)

**OTHER TRAINING WHICH SUPPORTS DEVELOPMENT OF KNOWLEDGE AND
SKILLS**
None

AD hiring authority and procurement of private services may not be used for this position.
* Online training at: _http://training.nwcg.gov/classes/i100.htm_

SECURITY SPECIALIST LEVEL 2 (SEC2)

REQUIRED TRAINING
Forest Protection Officer training
I-100 Introduction to Incident Command System*
IS-700 National Incident Management System (NIMS), An Introduction

CERTIFICATION
Annual recertification as a Forest Protection Officer

AUTHORITY
Not authorized or equipped to carry firearms, serve warrants, or conduct searches and seizures. Authority to enforce federal criminal laws and regulations.

REQUIRED EXPERIENCE
Satisfactory position performance as a Security Specialist Level 2 (SEC2)

PHYSICAL FITNESS LEVEL
None Required

OTHER POSITION ASSIGNMENTS THAT WILL MAINTAIN CURRENCY
None

OTHER TRAINING WHICH SUPPORTS DEVELOPMENT OF KNOWLEDGE AND SKILLS
None

AD hiring authority and procurement of private services may not be used for this position.
* Online training at: *http://training.nwcg.gov/classes/i100.htm*

SMALL ENGINE MECHANIC (SMEC)

REQUIRED TRAINING
I-100 Introduction to Incident Command System*
IS-700 National Incident Management System (NIMS), An Introduction
L-180 Human Factors on the Fireline
S-211 Portable Pumps and Water Use

CERTIFICATION
Fork Lift Operator

REQUIRED EXPERIENCE
Experience working within the National Cache System Small Engine Shop
        AND
Satisfactory performance as a Small Engine Mechanic (SMEC)

PHYSICAL FITNESS LEVEL
None Required

OTHER POSITION ASSIGNMENTS THAT WILL MAINTAIN CURRENCY
None

OTHER TRAINING WHICH SUPPORTS DEVELOPMENT OF KNOWLEDGE AND SKILLS
None

* Online training at: *http://training.nwcg.gov/classes/i100.htm*

STRIKE TEAM LEADER MILITARY (STLM)

REQUIRED TRAINING:
IS-700 National Incident Management System (NIMS), An Introduction
IS-800 National Response Framework (NRF), An Introduction
RT-130 Annual Fireline Safety Refresher

REQUIRED EXPERIENCE
Prior military experience is desirable, but not a required prerequisite
    AND
Satisfactory performance as a Strike Team Leader Crew (STCR)

PHYSICAL FITNESS LEVEL
Arduous

OTHER POSITION ASSIGNMENTS THAT WILL MAINTAIN CURRENCY
Strike Team Leader Crew (STCR)

OTHER TRAINING WHICH SUPPORTS DEVELOPMENT OF KNOWLEDGE AND SKILLS
None

TRACTOR PLOW OPERATOR INITIAL ATTACK (TPIA)

REQUIRED TRAINING
IS-700 National Incident Management System (NIMS), An Introduction
RT-130 Annual Fireline Safety Refresher
S-233 Tractor/Plow Boss
S-290 Intermediate Fire Behavior

CERTIFICATION
Local Tractor Plow Operator Certification

REQUIRED EXPERIENCE
Tractor Plow Operator (TPOP)
     AND
Satisfactory position performance as a Tractor Plow Operator Initial Attack (TPIA)

PHYSICAL FITNESS LEVEL
Moderate

OTHER POSITION ASSIGNMENTS THAT WILL MAINTAIN CURRENCY
Dozer Operator Initial Attack (DZIA)

OTHER TRAINING WHICH SUPPORTS DEVELOPMENT OF KNOWLEDGE AND SKILLS
None

TRACTOR PLOW OPERATOR (TPOP)

REQUIRED TRAINING
I-100 Introduction to Incident Command System*
IS-700 National Incident Management System (NIMS), An Introduction
RT-130 Annual Fireline Safety Refresher
S-130 Basic Firefighter
S-190 Introduction to Fire Behavior

CERTIFICATION
Local Tractor Plow Operator Certification

REQUIRED EXPERIENCE
Satisfactory performance as a Tractor Plow Operator (TPOP)

PHYSICAL FITNESS LEVEL
Moderate

OTHER POSITION ASSIGNMENTS THAT WILL MAINTAIN CURRENCY
Dozer Operator (DZOP)

OTHER TRAINING WHICH SUPPORTS DEVELOPMENT OF KNOWLEDGE AND SKILLS
None

* Online training at: *http://training.nwcg.gov/classes/i100.htm*

WAREHOUSE MATERIALS HANDLER (WHHR)

REQUIRED TRAINING
I-100 Introduction to Incident Command System*
IS-700 National Incident Management System (NIMS), An Introduction
L-180 Human Factors on the Fireline

CERTIFICATION
Hazmat Certification for 49 CFR

REQUIRED EXPERIENCE
Desirable to have experience working within the National Cache System
    AND
Experience working with the National Fire Equipment System (NFES)
    AND
Experience working with the National Interagency Cache Business System (ICBS)
    AND
Successful position performance as a Warehouse Materials Handler (WHHR)

PHYSICAL FITNESS LEVEL
None Required

OTHER POSITION ASSIGNMENTS THAT WILL MAINTAIN CURRENCY
None

OTHER TRAINING WHICH SUPPORTS DEVELOPMENT OF KNOWLEDGE AND SKILLS
None

Task Book available at: *http://www.nwcg.gov/pms/taskbook/taskbook.htm*
* Online training at: *http://training.nwcg.gov/classes/i100.htm*

WAREHOUSE MATERIALS HANDLER LEADER (WHLR)

REQUIRED TRAINING
I-100 Introduction to Incident Command System*
I-200 Basic Incident Command System
IS-700 National Incident Management System (NIMS), An Introduction

REQUIRED EXPERIENCE
Satisfactory performance as a Warehouse Materials Handler (WHHR)
     AND
Successful position performance as a Warehouse Materials Handler Leader (WHLR)

PHYSICAL FITNESS LEVEL
None Required

OTHER POSITION ASSIGNMENTS THAT WILL MAINTAIN CURRENCY
Warehouse Materials Handler (WHHR)

OTHER TRAINING WHICH SUPPORTS DEVELOPMENT OF KNOWLEDGE AND SKILLS
I-300 Intermediate Incident Command System
National Cache Demobilization Specialist Training

Task Book available at: *http://www.nwcg.gov/pms/taskbook/taskbook.htm*
* Online training at: *http://training.nwcg.gov/classes/i100.htm*

CHAPTER 3 – TRAINING DEVELOPMENT, LEADERSHIP REFRESHER TRAINING AND HISTORY

Effective Date: February 28, 2011, Updated 6/10/2011

Table of Contents

3.1 - TRAINING

Courses should be taken in an ascending order of complexity, based on successively higher levels of responsibility and skills in fire and aviation management.

3.11 - Instructor Qualifications, Training and Certification

Certification of instructor qualifications is the responsibility of the employing agency. Instructor qualifications, training and certification standards are contained within the NWCG Field Manager's Course Guide (FMCG), PMS 901-1. Refer to the FMCG for those standards which can be found at the following website:

http://training.nwcg.gov/sect_inst_certifications.htm

The Forest Service complies with the standards contained in the FMCG. In addition, Forest Service instructors shall:

1. Complete either the instructor requirements in the Field Manager's Course Guide for 200 and 300 level courses or the National Fire Protection Association (NFPA) 1041, Fire Service Instructor, which the Forest Service has identified as an equivalency course, for those courses in the Field Manager's Course Guide.

2. Meet NWCG or IAT Instructor qualification guidelines before delivering NWCG courses which have incorporated Interagency Aviation Training (IAT) materials, or any

A courses applicable to employees attempting to qualify for positions contained in FSFAQG.

Certification of lead and unit instructors for the "A" courses, who have not completed A-220 Train-the-Trainer, is the decision of the National Aviation Training Specialist, located in Boise, Idaho.

3.111 - Lead Instructor Requirements for "L" Courses

The L-180 Human Factors in the Wildland Fire Service and L-280 Followership to Leadership are training course packages available in the NWCG Publication Management System. Lead instructor requirements for these courses are defined in the NWCG Field Manager's Course Guide.

The L-380 Fireline Leadership, L-381 Incident Leadership, and L-480 Organizational Leadership in the Wildland Fire Service are training courses which do not have a standard NWCG course package available for lead instructors. Therefore, the U.S. Forest Service will use the following process to evaluate and approve providers, lead instructors, and the course packages which they develop to meet the intent and the criteria for these courses as established by the NWCG Leadership Subcommittee.

The course descriptions and design criteria are available at:

http://www.fireleadership.gov/courses/courses.html

Certification of prospective providers and their lead instructors submitting a new course package for L-380 Fireline Leadership, L-381 Incident Leadership, or L-480 Organizational Leadership in the Wildland Fire Service must be done by an evaluation team comprised of:

1. A team leader designated by the NWCG Leadership Subcommittee.

2. At least one additional evaluator from any of the agencies participating in NWCG.

Certification of new lead instructors for an existing approved L-380,

L-381, or L-480 course package shall be recommended by a peer evaluation from an individual who is currently certified as a lead instructor for that same course. The NWCG Leadership Subcommittee must then approve the recommendation. A format for these evaluation procedures is available upon request from the U.S. Forest Service representative to the NWCG Leadership Subcommittee.

http://www.fireleadership.gov/committee/committee.html

A list of approved providers and their respective certified lead instructors is available at:

http://www.fireleadership.gov/courses/courses.html

The L-580 Leadership is Action program is a series of continuing education offerings for senior level leaders in the wildland fire service. This program is managed by the L-580 Steering Committee. This steering committee is chartered under the NWCG Leadership Subcommittee and works in partnership with the National Adavanced Fire and Resource Institute (NAFRI). The U.S. Forest Service will recognize only offerings approved by this steering committee as L-580 events.

3.112 - Credit for Teaching National Wildfire Coordinating Group Training Courses

When serving as an instructor, a unit or adjunct instructor may receive credit for attending a course, provided the following conditions are met:

1. Prior agreement with the lead instructor and course coordinator is reached for the unit or adjunct instructor to complete the course.

2. The unit or adjunct instructor completing the course must meet all course prerequisites.

3. The entire training session must be attended, with adequate participation in class or small group activities.

4. Any required pre-course work material must be successfully completed.

5. All unit and course exams must be successfully completed.

Once the above conditions have been met, the Lead Instructor may issue a course completion certificate to the unit or adjunct instructor.

3.12 - Course Delivery Standards

The Forest Service shall comply with the course delivery standards contained within the NWCG Field Manager's Course Guide. Refer to the FMCG for those standards which can be found at the following website:

http://www.nwcg.gov/pms/training/fmcg.pdf

3.13 - Forest Service Fireline Safety Refresher Training

Annual Fireline Safety Refresher Training is required for all positions as identified in FSFAQG and the *Wildland Fire Qualifications System Guide* (NWCG 310-1). Forest Service has extended this requirement to 13 months. Annual Fireline Safety Refresher Training is provided in order to recognize hazards and mitigate risk, maintain safe practices and to reduce accidents and near misses. The intent of the annual fireline safety refresher training is to focus suppression and prescribed fire personnel on operations and decision making issues related to incident safety.

Annual Fireline Safety Refresher Training must include the following core topics:	
Core Topics	**Examples**
Avoiding Entrapments	Use training and reference materials to study the risk management process as identified in the Incident Response Pocket Guide as appropriate to the participants, e.g., LCES, Standard Firefighting Orders, Eighteen Watch Out Situations, Wildfire Decision Support System (WFDSS) direction, Fire Management Plan priorities, etc.
Current Issues	Review and discuss identified "hot topics" as found on the current Wildland Fire Safety Training Annual Refresher (WFSTAR) website. Review forecasts and assessments for the upcoming fire season and discuss implications for firefighter safety.
Fire Shelter	Review and discuss last resort survival including escape and shelter deployment site selection. Conduct "hands-on" fire shelter inspections. Practice shelter deployments in applicable crew/module 18 configurations.
Other Hazards and Safety Issues	Choose additional hazard and safety subjects, which may include SAFENET, current safety alerts, site/unit specific safety issues and hazards.

These core topics must be sufficiently covered to ensure that personnel are aware of safety concerns and procedures and can demonstrate proficiency in fire shelter deployment.

The Forest Service has <u>No</u> minimum refresher training hour requirements. Core topics shown above should reflect the quality of the material used and not the quantity.

Further guidance can be found at the following links:

Interagency Standards for Fire and Fire Aviation Operations (Red Book)

http://www.nifc.gov/policies/red_book/2011/Ch13.pdf

Wildland Fire Safety Training Annual Refresher (WFSTAR)

http://www.nifc.gov/wfstar/index.htm

3.2 - EQUIVALENCY COURSES

Equivalency courses are classes that are adequate substitutes for National Wildfire Coordinating Group (NWCG) approved curriculum and that are approved by the Washington Office, Branch Chief for Fire Training. Approved equivalency courses are listed in exhibit 01.

1. <u>Process to Evaluate and Establish Equivalency Courses</u>. The appropriate Regional Training Working Team or steering committee shall identify the need for an equivalency analysis of a specific course. The committee shall assign an evaluation team (see para. 2 regarding the team composition) to conduct the analysis, document their findings, and submit recommendations through agency channels to the Washington Office, Fire and Aviation Management, Branch Chief for Fire Training for an equivalency review.

a. If the Branch Chief for Fire Training determines that the equivalency course analysis is sufficient and the proposed course meets the NWCG certified course standards, the Branch Chief shall recognize the course as equivalent.

b. The Branch Chief may also recommend acceptance of the equivalency course(s) to the NWCG Operations and Workforce Development Committee (OWDC).

2. Evaluation Team Composition. The evaluation team shall be comprised of a minimum of three of the following members, including: lead instructor, cadre member, and course developer or subject matter expert for the respective NWCG course. The evaluators shall be individuals either who have been involved within the past 3 years with instructing the NWCG course, or who are familiar with the course development and revision process.

3. Equivalency Courses. The Branch Chief for Fire Training has determined that the courses listed in exhibit 01 are equivalent to the identified NWCG course. Persons who have successfully completed the identified equivalency course do not need to attend the corresponding NWCG course.

3.2 - Exhibit 01

List of Approved Equivalency Courses

NWCG Approved Curriculum	Approved Equivalency Course(s) and Experience
D-312 Aircraft Dispatcher	Bureau of Land Management Aviation Dispatcher This equivalency is retroactive for all individuals who previously completed the BLM Aviation Dispatcher course. Competency for D-312 should be granted in the employee's IQCS file.
L-180 - Human Factors on the Fireline	S-130 - Firefighter Training (2003 version). The 2003 version of S-130 incorporates L-180 into the course package. Individuals completing the 2003 version of S-130 should be given course completion certificates for both S-130 and L-180, both courses should be entered into the Incident Qualification and Certification System (IQCS).
L-380 - Fireline Leadership This course was also delivered under the following titles in 2001-2003: FMO Leadership Workshop Leading in Fire Management	Employees who completed L-380 can be granted course competency in IQCS for the L-180 Human Factors on the Fireline and L-280 Followership to Leadership courses with the following justification statement: "Employee's Name" completed L-380 prior to L-180 and L-280. Course competency has been granted for these courses.
L-381 - Incident Leadership	Employees who completed L-381 can be granted course competency in IQCS for L-180 Human Factors on the Fireline.
M-410 - Facilitative Instructor	National Fire Protection Association (NFPA) 1041, Fire Service Instructor I, with proficiency as set out in the NWCG Field Manager's Course Guide. National Fire Protection Association (NFPA) 1a and 1b (must complete both courses), with proficiency as set out in the NWCG Field Manager's Course Guide.
M-410 Facilitative Instructor and NFPA 1041	A-220 Train-the-Trainer (for delivery of the "A" course curriculum only)

3.2 - Exhibit 01--Continued

S-110 - Basic Wildland Fire Orientation	Experience in operations positions on an incident. S-110 is designed for non-operations personnel slated for a first on-incident assignment. Many of the Technical Specialist positions listed in chapter 20 reflect S-110 as required training. However, if the incumbent has had fireline experience or previous incident experience, S-110 is not required. In these instances, course competency for S-110 should be granted in IQCS with a justification statement explaining that the individual has previous incident experience.
S-336 - Fire Suppression Tactics	Successful completion of either: S-230 (1996 version) - Single Resource Boss AND S-215 - Fire Operations in the Urban Interface OR S-330 - Task Force/Strike Team Leader AND S-215 - Fire Operations in the Urban Interface
S-580 - Advanced Fire Use Applications	Managing Wildland Fire for Resource Benefits (offered in Region 1). Applies only to Strategic Operational Planner qualifications.
RX-310 - Introduction to Fire Effects	Successful completion of Technical Fire Management.

3.21 - Historical Information for Equivalency Courses

In 2002 the Forest Service developed an equivalency process for evaluating courses which adequately substitute for NWCG courses. The following courses were removed from the NWCG curriculum, but are provided for historical documentation in an employee's master file record (FSH 5109.17, sec. 22.1, Record Keeping).

3.21 - Exhibit 01

List of Courses Removed From NWCG Curriculum

NWCG Approved Curriculum	Approved Equivalency Course(s)
S-201/S-281 - Supervisory Concepts and Techniques (Prior to 10/1/2003) Note: Removed from the NWCG curriculum in October 2003. Credit should not be given for the course after 9/30/2003.	Forest Service Corporate Training Practical Leadership Skills for New First-Line Supervisors. Note: The Forest service does not recognize S-201/S-281 as equivalent to L-280 Followership to Leadership.
S-301/S-381 - Leadership and Organizational Development Note: Removed from Forest Service recognition in July 2003. Removed from the NWCG curriculum in October 2004.	L-380 Fireline Leadership. This course was also delivered under the following titles in 2001-2003: FMO Leadership Workshop Leading in Fire Management
Interagency Aviation Management and Safety (IAMS)	The following modules may be offered at the Aviation Centered Education (ACE) or Regional Workshops and are equivalent to the NWCG IAMS course for identified positions: Supervisory Dispatcher (EDSP): A-101 Basic Aircraft Safety A-104 Aircraft Capabilities and Limitations A-106 Aircraft Mishap Reporting A-109 Aircraft Radio Use A-112 Mission Planning and Flight Request Process A-202 Interagency Aviation Organizations A-203 Basic Airspace A-206 Aviation Acquisition/Procurement I A-207 Aircraft Dispatching A-302 Personal Responsibility and Liability A-303 Human Factors in Aviation A-305 Risk Management A-307 Aviation Policy and Regulations II Helicopter Manager: A-101 Aviation Safety A-103 Helicopter Safety A-104 Overview of Aircraft Capabilities and Limitations A-105 Aviation Life Support and Equipment A-106 Aviation Mishap Reporting A-107 Aviation Policy and Regulations A-108 Pre-Flight Checklist and Briefing/Debriefing A-112 Mission Planning and Flight Request Process A-113 Crash Survival

3.21 - Exhibit 01--Continued

Interagency Aviation Management and Safety (IAMS)	Air Operations Branch Director (AOBD), Air Tactical Group Supervisor (ATGS), Air Support Group Supervisor (ASGS), Air Tanker/Fixed Wing Coordinator (ATCO), Ramp Manager (RAMP), and Fixed Wing Base Manager (FWBM): A-101 Basic Aircraft Safety A-102 Fixed Wing Safety A-103 Helicopter Safety A-105 Aviation Life Support Equipment A-106 Aircraft Mishap Reporting A-107 Aviation Policy and Regulations 1 A-109 Aircraft Radio Use A-112 Mission Planning and Flight Request Process A-113 Crash Survival A-202 Interagency Aviation Organizations A-203 Basic Airspace A-204 Aircraft Capabilities and Limitations A-206 Aviation Acquisition/Procurement I A-311 Aviation Planning A-301 Implementing Aviation Safety and Accident Programs A-302 Personal Responsibility and Liability A-303 Human Factors in Aviation A-305 Risk Management

More information on NWCG course History is located at:

http://training.nwcg.gov/sect_training_curriculum.htm

3.22 – Interchangeable Courses

NWCG has approved interchangeable courses. A description of the interchangeable course guidelines and approved interchangeable courses can be found in the Field Manager's Course Guide: *http://www.nwcg.gov/pms/training/fmcg.pdf*

The ICS-100 and ICS-200 courses accessible in AgLearn have not been deemed to be interchangeable.

For positions in this handbook which require I-100 and I-200, equivalency can be met by successfully completing one of the approved interchangeable courses listed in the Field Manager's Course Guide.

3.23 - Course Development History

The NWCG Course Development and Standards Division provide historical information on course development which includes course numbering and course title changes. Information is available on the following website, under "Course Development/Revision Status":

http://training.nwcg.gov/sect_training_curriculum.htm

The following exhibit is an addendum to the information contained in the "Curriculum Status" website, and it includes dates of certification and removal of courses from the NWCG curriculum:

3.23 - Exhibit 01
Addendum to Curriculum Status Website

Development History	NWCG Approved Curriculum - Equivalent Course(s)
I-220 - Basic Incident Command System, certified in 1983, obsolete after November 1993	I-100 - Introduction to Incident Command System, available Feb. 1994 I-200 - Basic Incident Command System, available Feb. 1994
I-375 - Air Support Group Supervisor, certified in 1986, obsolete after December 1996	J-375 - Air Support Group Supervisor, certified in 1997
P-151 - Wildfire Origin and Cause Determination, obsolete after 2003	FI - 210 - Wildland Fire Origin and Cause Determination, certified in 2005
Prescribed Fire Management, obsolete after 2000	RX-300 - Prescribed Fire Burn Boss, certified in 2000
Managing Fire Effects, obsolete after 1995	RX-310/RX-340 - Introduction to Fire Effects, certified in 1995
S-213 - Tractor Use/Tractor Boss, certified in 1979; obsolete after November 1995	S-233 - Tractor/Plow Boss, Single Resource, certified in 1995
S-215 - Firing Equipment/Firing Boss, certified in 1979; obsolete after December 1991	S-234 - Ignition Operations, certified in 1991
S-261 - Personnel Timerecorder, S-262 - Equipment Timerecorder; S-263 - Claims Specialist, S-264 - Compensation for Injury Specialist; S-266 - Commissary Manager all courses were certified in 1988; obsolete after December 1996	S-261 - Applied Interagency Incident Business Management, certified in 1999
S-390 - Intermediate Wildland Fire Behavior, certified in 1981, obsolete after November 1993	S-290 - Intermediate Wildland Fire Behavior, available in February 1994 S-390 - Introduction to Wildland Fire Behavior Calculations, available in February 1994
I-330 - Task Force/Strike Team Leader; I-333 - Strike Team Leader, Crew; I-334 - Strike Team Leader, Engine; I-335 - Strike Team Leader, Dozer; certified in 1984 and obsolete after September 1995	S-330 - Task Force/Strike Team Leader, certified in 1996
I-339 - Division/Group Supervisor, certified in 1985, obsolete after September 1995	S-339 - Division/Group Supervisor, certified in 1997
I-362 - Cost Unit Leader; I-363 - Compensation/Claims Unit Leader; I-365 - Time Unit Leader; I-368 - Procurement Unit Leader; certified in 1987, obsolete after 2000	S-360 - Finance/Administration Unit Leader, certified in 2000
I-271 - Helibase Manager, certified in 1986, obsolete after December 1996	S-371 - Helibase Manager, certified in 1997
I-403 - Information Officer, certified in 1988, obsolete after 2001	S-403 - Information Officer, certified in 2001
I-401 - Safety Officer, certified in 1986	S-404 - Safety Officer, certified in 2002

3.3 - DEVELOPMENT

3.31 - Supervisory Development/Leadership

If an employee has not yet completed the Forest Service second 40 hours of supervision training as required in FSH 6109.13 (the Branch of Corporate Training course entitled, "Leadership Skills for Experienced Managers and Supervisors"), then completion of L-380 Fireline Leadership, or L-381 Incident Leadership, or L-480 Organizational Leadership in the Wildland Fire Service have been deemed as equivalent courses. However, completion of "Leadership Skills for Experienced Managers and Supervisors" has not been deemed to be an equivalent course to L-380, L-381, or L-480.

The contents and format of the second 40 hours of supervisor training are determined by the supervisor or management of the local unit and should be adapted to the needs of the employee's job.

Provided that the employee's supervisor approves L-380 or L-381, or L-480 as part of the employee's Individual Development Plan (IDP) as well as recorded in the Incident Qualification and Certification System (IQCS) Incident Responder Development Plan (IRDP), these training courses should meet the supervisory training requirement. Further direction regarding supervision training is found in FSH 6109.13.

FIRE AND AVIATION
QUALIFICATIONS GUIDE

CHAPTER 4 - FIRE AND AVIATION MANAGEMENT POSITION COMPETENCIES
HOTSHOT, EXCLUSIVE USE HELITACK AND IFPM/FS-FPM

Effective Date: **February 28, 2011**

Table of Contents

4.1 - FIRE AND AVIATION POSITION COMPETENCIES

The Federal Fire and Aviation Leadership Council has developed the Interagency Fire Program Management Qualifications Standards and Guide, which contains minimum qualification standards for fire and aviation management positions. The Forest Service has determined that in addition to the minimum qualification standards contained in IFPM and FS-FPM, the following competencies are required to reach full performance in each associated position listed in sections 4.11, 4.12 and 4.13.

4.11 - Hotshot Position Competencies

HOTSHOT SUPERINTENDENT

REQUIRED TRAINING:	M-410 Facilitative Instructor or equivalent I-300 Intermediate ICS L-380 Fireline Leadership S-200 Initial Attack IC S-330 Task Force/Strike Team Leader S-390 Intro to Fire Behavior Calculations
REQUIRED DEVELOPMENTAL TRAINING (must be obtained within the first year):	N/A
CERTIFICATION:	RT-130 Annual Fireline Safety Refresher Annual Operational Preparedness Training
PREREQUISITE EXPERIENCE:	IFPM Minimum Qualification Standards for IHCS (see IFPM Crosswalk located at http://www.fs.fed.us/fire/management/ifpm/crosswalk.pdf)
PHYSICAL FITNESS:	Arduous

*Desired fitness goals also include completing:
1.5-mile run in a time of 10:35 or less
                    AND
Forty sit-ups in sixty seconds
                    AND
Twenty-five push-ups in sixty seconds
                    AND
Four chin-ups (>170 lbs. body weight)
Five chin-ups (135-170 lbs. body weight)
Six chin-ups (110-134 lbs. body weight)
Seven chin-ups (<110 lbs. body weight)

* Reference "Standards for Interagency Hotshot Crew Operations" (FSM 5108)

HOTSHOT ASSISTANT SUPERINTENDENT/CAPTAIN

REQUIRED TRAINING:

I-300 Intermediate ICS
L-380 Fireline Leadership
S-200 Initial Attack IC
S-330 Task Force/Strike Team Leader
S-390 Intro to Fire Behavior Calculations
AND
Hotshot Squad Leader Required Training

REQUIRED DEVELOPMENTAL TRAINING (must be obtained within the first year):

M-410 Facilitative Instructor or equivalent

CERTIFICATION:

RT-130 Annual Fireline Safety Refresher
Annual Operational Preparedness Training

PREREQUISITE EXPERIENCE:

FS-FPM Minimum Qualification Standards for IHCAS
(see FS-FPM Crosswalk located at
http://www.fs.fed.us/fire/management/ifpm/crosswalk.pdf)

PHYSICAL FITNESS:

Arduous

*Desired fitness goals also include completing:
1.5-mile run in a time of 10:35 or less
AND
Forty sit-ups in sixty seconds
AND
Twenty-five push-ups in sixty seconds
AND
Four chin-ups (>170 lbs. body weight)
Five chin-ups (135-170 lbs. body weight)
Six chin-ups (110-134 lbs. body weight)
Seven chin-ups (<110 lbs. body weight)

* Reference "Standards for Interagency Hotshot Crew Operations" (FSM 5108)

HOTSHOT SQUAD LEADER

REQUIRED TRAINING: Hotshot Senior Firefighter Required Training

REQUIRED DEVELOPMENTAL   I-200 Basic ICS
TRAINING (must be obtained   L-280 Followership to Leadership
within the first year):   S-215 Fire Operations in the Wildland/Urban Interface
                                                      S-230 Crew Boss (Single Resource)
                                                      S-234 Ignition Operations
                                                      S-260 Interagency Incident Business Management

CERTIFICATION: RT-130 Annual Fireline Safety Refresher
                                                      Annual Operational Preparedness Training

PREREQUISITE EXPERIENCE: FS-FPM Minimum Qualification Standards for IHCSQL
                                                      (see FS-FPM Crosswalk located at
                                                      http://www.fs.fed.us/fire/management/ifpm/crosswalk.pdf)

PHYSICAL FITNESS: Arduous

                                                      *Desired fitness goals also include completing:
                                                      1.5-mile run in a time of 10:35 or less
                                                               AND
                                                      Forty sit-ups in sixty seconds
                                                               AND
                                                      Twenty-five push-ups in sixty seconds
                                                               AND
                                                      Four chin-ups (>170 lbs. body weight)
                                                      Five chin-ups (135-170 lbs. body weight)
                                                      Six chin-ups (110-134 lbs. body weight)
                                                      Seven chin-ups (<110 lbs. body weight)

* Reference "Standards for Interagency Hotshot Crew Operations" (FSM 5108)

HOTSHOT SENIOR FIREFIGHTER

REQUIRED TRAINING:	S-131 Firefighter Type 1 S-133 Look Up, Look Down, Look Around S-211 Portable Pumps S-212 Wildland Power Saws S-290 Intermediate Fire Behavior **AND** Hotshot Crewmember Required Training
REQUIRED DEVELOPMENTAL TRAINING (must be obtained within the first year):	S-270 Basic Air Operations
CERTIFICATION:	RT-130 Annual Fireline Safety Refresher Annual Operational Preparedness Training
PREREQUISITE EXPERIENCE:	IFPM Minimum Qualification Standards for SFF (see IFPM Crosswalk located at http://www.fs.fed.us/fire/management/ifpm/crosswalk.pdf)
PHYSICAL FITNESS:	Arduous *Desired fitness goals also include completing: 1.5-mile run in a time of 10:35 or less **AND** Forty sit-ups in sixty seconds **AND** Twenty-five push-ups in sixty seconds **AND** Four chin-ups (>170 lbs. body weight) Five chin-ups (135-170 lbs. body weight) Six chin-ups (110-134 lbs. body weight) Seven chin-ups (<110 lbs. body weight)

* Reference "Standards for Interagency Hotshot Crew Operations" (FSM 5108)

HOTSHOT CREWMEMBER

REQUIRED TRAINING:
I-100 Intro to ICS
L-180 Human Factors
S-130 Firefighter Training
S-190 Intro to fire Behavior

RECOMMENDED TRAINING:
S-290 Intermediate Fire Behavior

CERTIFICATION:
RT-130 Annual Fireline Safety Refresher
Annual Operational Preparedness Training

PREREQUISITE EXPERIENCE:
Firefighter Type 2

PHYSICAL FITNESS:
Arduous

*Desired fitness goals also include completing:
1.5-mile run in a time of 10:35 or less
AND
Forty sit-ups in sixty seconds
AND
Twenty-five push-ups in sixty seconds
AND
Four chin-ups (>170 lbs. body weight)
Five chin-ups (135-170 lbs. body weight)
Six chin-ups (110-134 lbs. body weight)
Seven chin-ups (<110 lbs. body weight)

* Reference "Standards for Interagency Hotshot Crew Operations" (FSM 5108)

4.12 - Exclusive Use Fire Helicopter Position Competencies

EXCLUSIVE-USE FIRE HELICOPTER CREW SUPERVISOR

REQUIRED TRAINING:	A-200 Mishap Review A-310 Overview of Crew Resource Management Aviation Contract Administration Course according to level of responsibility
REQUIRED DEVELOPMENTAL TRAINING (must be obtained within the first year):	I-300 Intermediate Incident Command System
CERTIFICATION:	RT-130 Annual Fireline Safety Refresher RT-372 Helicopter Manager Workshop (Triennial)
PREREQUISITE EXPERIENCE:	One season's experience as Exclusive-Use Assistant Helicopter Crew Supervisor **AND** IFPM Minimum Qualification Standards for HMGR (see IFPM Crosswalk located at http://www.fs.fed.us/fire/management/ifpm/crosswalk.pdf)
PHYSICAL FITNESS:	Arduous Desired fitness goals also include completing: 1.5-mile run in a time of 10:35 or less **AND** Forty sit-ups in sixty seconds **AND** Twenty-five push-ups in sixty seconds **AND** Four chin-ups (>170 lbs. body weight) Five chin-ups (135-170 lbs. body weight) Six chin-ups (110-134 lbs. body weight) Seven chin-ups (<110 lbs. body weight)

EXCLUSIVE-USE ASSISTANT FIRE HELICOPTER CREW SUPERVISOR

REQUIRED TRAINING:

S-372 Helicopter Management
S-371 Helibase Manager
Aviation Contract Administration
Crew Resource Management

REQUIRED DEVELOPMENTAL TRAINING (must be obtained within the first year):

CERTIFICATION:

RT-130 Annual Fireline Safety Refresher
RT-372 Helicopter Manager Workshop (Triennial)

PREREQUISITE EXPERIENCE:

One season's experience as Exclusive Use Helicopter Squad Leader
AND
Experience in Aviation Contract Administration
AND
FS-FPM Minimum Qualification Standards for FHACS (see FS-FPM Crosswalk located at
http://www.fs.fed.us/fire/management/ifpm/crosswalk.pdf)

PHYSICAL FITNESS:

Arduous
Desired fitness goals also include completing:
1.5-mile run in a time of 10:35 or less
AND
Forty sit-ups in sixty seconds
AND
Twenty-five push-ups in sixty seconds
AND
Four chin-ups (>170 lbs. body weight)
Five chin-ups (135-170 lbs. body weight)
Six chin-ups (110-134 lbs. body weight)
Seven chin-ups (<110 lbs. body weight)

EXCLUSIVE-USE FIRE HELICOPTER SQUAD LEADER

<u>REQUIRED TRAINING:</u>	S-131 Advanced Firefighter S-133 Look Up, Look Down, Look Around S-211 Portable Pumps and Water Use S-212 Wildfire Power Saws
<u>REQUIRED DEVELOPMENTAL</u> TRAINING (must be obtained within the first year):	S-372 Helicopter Management
<u>CERTIFICATION:</u>	RT-130 Annual Fireline Safety Refresher S-271 Annual Helicopter Crewmember Refresher (must complete course or approved refresher or serve as an instructor)
<u>PREREQUISITE EXPERIENCE:</u>	One season experience as a fire Helicopter Crewmember **AND** FS-FPM Minimum Qualification Standards for FHSQL (see FS-FPM Crosswalk located at http://www.fs.fed.us/fire/management/ifpm/crosswalk.pdf)
<u>PHYSICAL FITNESS:</u>	Arduous Desired fitness goals also include completing: 1.5-mile run in a time of 10:35 or less **AND** Forty sit-ups in sixty seconds **AND** Twenty-five push-ups in sixty seconds **AND** Four chin-ups (>170 lbs. body weight) Five chin-ups (135-170 lbs. body weight) Six chin-ups (110-134 lbs. body weight) Seven chin-ups (<110 lbs. body weight)

EXCLUSIVE-USE FIRE HELICOPTER SENIOR FIREFIGHTER

REQUIRED TRAINING:

S-131 Firefighter Type 1
S-133 Look Up, Look Down, Look Around
S-211 Portable Pumps
S-212 Wildland Power Saws
S-290 Intermediate Fire Behavior

REQUIRED DEVELOPMENTAL
TRAINING (must be obtained
within the first year):

S-270 Basic Air Operations

CERTIFICATION:

RT-130 Annual Fireline Safety Refresher
Annual Operational Preparedness Training

PREREQUISITE EXPERIENCE:

IFPM Minimum Qualification Standards for SFF
(see IFPM Crosswalk located at
http://www.fs.fed.us/fire/management/ifpm/crosswalk.pdf)

PHYSICAL FITNESS:

Arduous

Desired fitness goals also include completing:
1.5-mile run in a time of 10:35 or less
AND
Forty sit-ups in sixty seconds
AND
Twenty-five push-ups in sixty seconds
AND
Four chin-ups (>170 lbs. body weight)
Five chin-ups (135-170 lbs. body weight)
Six chin-ups (110-134 lbs. body weight)
Seven chin-ups (<110 lbs. body weight)

EXCLUSIVE-USE FIRE HELICOPTER CREWMEMBER

REQUIRED TRAINING:	I-100 Introduction to Incident Command System
	S-190 Introduction to Wildland Fire Behavior
	S-130 Firefighting Training
REQUIRED DEVELOPMENTAL TRAINING (must be obtained within the first year):	S-271 Interagency Helicopter Training
	S-290 Intermediate Fire Behavior
CERTIFICATION:	RT-130 Annual Fireline Safety Refresher
	S-271 Annual Helicopter Crewmember Refresher (must complete course or approved refresher or serve as an instructor)
PREREQUISITE EXPERIENCE:	One Season as a Firefighter Type 2

PHYSICAL FITNESS:

Arduous

Desired fitness goals also include completing:
1.5-mile run in a time of 10:35 or less

AND

Forty sit-ups in sixty seconds

AND

Twenty-five push-ups in sixty seconds

AND

Four chin-ups (>170 lbs. body weight)
Five chin-ups (135-170 lbs. body weight)
Six chin-ups (110-134 lbs. body weight)
Seven chin-ups (<110 lbs. body weight)

EXCLUSIVE-USE RESTRICTED/LIMITED FIRE HELICOPTER MANAGER

REQUIRED TRAINING:	S-372 Helicopter Management Aviation Contract Administration Course according to level of responsibility
REQUIRED DEVELOPMENTAL TRAINING (must be obtained within the first year):	Crew Resource Management
CERTIFICATION:	RT-130 Annual Fireline Safety Refresher RT-372 Helicopter Management (Triennial)
PREREQUISITE EXPERIENCE:	Helicopter Manager (HMGB) **AND** Helibase Manager 2 Trainee (HEB2-T)
PHYSICAL FITNESS:	Moderate

4.13 – Forest Service Implementation of Interagency Fire Program Management (IFPM) and Forest Service Fire Program Management (FS-FPM).

On October 1, 2004 the Forest Service, in conjunction with the Department of Interior wildland agencies, implemented the Interagency Fire Program Management (IFPM) Standard and Guide. The intent of IFPM is to establish minimum qualification standards for 13 key fire management positions to ensure a minimum level of NWCG qualifications, specialized experience, and training was required for each position. All Fire and Aviation Management employees that occupy one of the IFPM positions were required to meet all of the position standards as of October 1, 2010 and thereafter.

In 2008, the Forest Service created an agency addendum to IFPM called Forest Service Fire Program Management (FS-FPM) to establish the minimum qualification standards for key fire management positions that were subordinate to one of the IFPM positions, or were located on a subunit (i.e. Ranger District). The implementation period for FS-FPM is scheduled to conclude on October 1, 2013, and all Fire and Aviation Management employees that occupy one of the FS-FPM positions are required to meet the position standards on that date and thereafter.

The following is a summary of all the IFPM and FS-FPM fire and aviation management positions recognized by the Forest Service.

<u>4.13 - Exhibit 01</u>

IFPM AND FS-FPM STANDARD POSITIONS		
IFPM Category	**FS Positions in IFPM**	**FS Positions in FS-FPM**
Unit Fire Program Manager	Forest FMO	Forest AFMO District or Zone FMO
Wildland Fire Operations Specialist	N/A	District or Zone AFMO T2 Handcrew Supervisors IA Module Leaders Station Managers
Prescribed Fire and Fuels Specialists	Forest Fuels Planner	District or Zone Fuels Specialists
Supervisory Engine Operator	Engine Captains Asst Engine Captains	
Engine Module Supervisor	Engine Captains	
IHC Superintendent	IHC Superintendent	IHC Asst Superintendents IHC Squad Leaders
Helicopter Manager	Helitack Crew Supervisor	Helitack Asst Crew Supervisors Helitack Squad Leaders
Senior Firefighter	Senior Firefighter	
Center Manager	Dispatch Center Manager	
Lead IA Dispatcher. Assistant Center Manager	Forest Dispatcher Asst Center Manager	
Initial Attack Dispatcher	Initial Attack Dispatcher	
Geographic Area Fire Program Manager	Regional Office	
National Fire Program Manager	National Office	

The Forest Service IFPM/FS-FPM Standard Position Description Crosswalk and implementation

procedures for IFPM and FS-FPM can be located at:

http://www.fs.fed.us/fire/management/ifpm/index.html

It is the responsibility of each Forest Service unit to ensure all encumbered FS-FPM

employees meet their minimum qualification standards by October 1, 2013. This includes

prioritizing training and associated costs as well as providing employees with work time to

complete the training.